CHASES AND ESCAPES

Frontispiece - Screenshot from the author's computer display,* captured while playing the World War II first-person shooter *Call of Duty*^{TM}:*United Offensive*^{TM}, showing what a defending tail machine gunner on a B-17 bomber saw in his gun sight when under attack over Europe by German fighters flying on 'pure pursuit' trajectories (see figure 2.6.3). The B-17 used an optical gunsight to help the gunner aim his two .50-caliber machine guns so their bullets would arrive at where the fighter "would be," rather than "where it was." Some gunners took a more direct approach; as one B-24 gunner who flew against the Japanese in the Pacific war put it, "My own feeling about it was, you just filled the sky full of lead and let them run into it." In this book, of course, we prefer to be just a *bit* more analytical than *that*! During World War II many classified theoretical studies of pursuit trajectories in air-to-air combat were made by the Applied Mathematics Panel (AMP), which was part of the U.S. government National Defense Research Committee (NDRC). The NDRC was created by presidential order in 1940, and the AMP was formed in 1942, again by written order of President Franklin D. Roosevelt. You can find an example of AMP's pursuit work, in the open literature, in Handelman (1949), and more on AMP's work, in general, in Rees (1980).

* For readers who also enjoy computer gaming, the author's old system was a DELL Inspiron 5150 laptop with 512 megabytes of RAM, a 3.06 GHz Pentium 4 Processor, and an Nvidia FX5200 graphics card with 64 megabytes of video memory, all running under Windows XP-Pro/DirectX 9.0c. By the standards of 2006, this was a low-powered gaming machine. The author now performs advanced gaming "research" on an Xbox360.

CHASES AND ESCAPES

The Mathematics of
Pursuit and Evasion

Paul J. Nahin

Princeton University Press
Princeton and Oxford

Copyright © 2007 by Princeton University Press
Published by Princeton University Press,
41 William Street,
Princeton, New Jersey 08540
In the United Kingdom:
Princeton University Press,
3 Market Place, Woodstock,
Oxfordshire OX20 1SY

All Rights Reserved

Library of Congress Cataloging-in-Publication Data

Nahin, Paul J.
Chases and escapes: the mathematics of pursuit and evasion/Paul J. Nahin.
p.   cm.
Includes bibliographical references and index.
ISBN-13: 978-0-691-12514-5 (acid-free paper)
ISBN-10: 0-691-12514-7 (acid-free paper)
1. Differential games. I. Title.

QA272.N34 2007
519.3′2–dc22 2006051392

British Library Cataloging-in-Publication Data
is available

This book has been composed in ITC New Baskerville

Printed on acid-free paper. ∞

press.princeton.edu

Printed in the United States of America

10 9 8 7 6 5 4 3 2

For Patricia Ann

Who chased whom?

It was forty-five years ago,

so who's to know?

Today it wouldn't matter,

we're **both** too slow!

If E is an evader with speed 1 and P is a pursuer with speed $s > 1$, then "Of course P can catch E [no matter what E does], at least by going to his [E's] initial position and [simply] following his path."

—from Isbell (1967), showing that not all pursuit problems have complicated answers

Contents

What You Need to Know

to Read This Book (and How I Learned What I Needed to Know to Write It)

If you have had the usual first two years of undergraduate mathematics taken by all mathematics, engineering, and physical science students (calculus and ordinary differential equations) then you are good-to-go for what follows. Geometry and trigonometry are often used, too, of course, but if you understand what a differential equation is then I think I can safely assume that the Pythagorean theorem and the law of cosines will not trip you up. Differential equations are the signature mathematics of this book (although we *will* use a computer a lot, too, and in one section even some very elementary probability arguments as well). If you can get your hands on it, look up the science fantasy short story "Those Who Can, Do" by Bob Kurosaka, which appeared in the January 1965 issue of *Fantasy and Science Fiction Magazine* (it was reprinted in *The Year's Best SF: 11th Annual Edition*, Delacorte 1966, Judith Merril, editor). Kurosaka — who was then an instructor in mathematics at Massachusetts Bay Community College — said he wrote his story in response to a student's challenge one day in his differential equations class, demanding to know "What's this stuff good for anyway?" The rest of this book should provide you with numerous answers to that question (which I heard at least a thousand times myself during thirty-one years of college teaching at the University of

New Hampshire, the University of Virginia, the Naval Postgraduate School, and Harvey Mudd College).

With the exception of a few isolated calculations (particularly in appendix D), there is nothing original in this book, by which I mean that nearly all of the mathematical results derived here have previously appeared in print, somewhere, sometime, in some cases *centuries* ago. Over a period of several years I gradually assembled numerous journal papers and identified textbook problems of particular historical and/or mathematical interest; the ones I found most valuable to my writing are listed in the bibliography at the end of the book. When I refer to one of those sources I have used the simple style "author (year)," for example, as in Smith (1917), where you'll find the following comment on an "interesting law of book writers — that most of them will steal from one another without the least scruple if they can thinly veil the theft." I admit that I have freely "stolen" from my sources, but I have also tried very hard *not* to veil my thefts. The bibliography lists the "victims" whose pages I picked with pleasure.

CHASES AND ESCAPES

Introduction

"Chases and escapes" are activities that might be said to define human existence, no matter where or when, ranging over the entire spectrum of violence from the romantic pursuit of a future spouse to the military pursuit of a target to be destroyed. Young children are introduced to both chases and escapes when they learn the simple rules of the game of tag: played by at least two, one player is designated as *it*, who then chases after all the others, who, of course, attempt to escape. When the *it* player manages to catch one of the evaders he or she yells "Tag! You're *it*!," that evader and the pursuer change roles, and the game continues until all drop from exhaustion. Parents love tag! The related game of hide-and-seek, also of this genre, is equally attractive to young children (and to their parents). And surely all readers can remember the fun of watching the old Looney Tunes Road Runner/Wile E. Coyote chase-and-escape cartoons. Hollywood's interest in chases dates, in fact, at least as far back as the 1914 serial *The Perils of Pauline*, with countless imaginative escapes by the heroine from her dastardly guardian's attempts to do her in.

As we grow older we leave hide-and-seek, tag, and cartoons behind, but there remains, for most people, a deep interest in pursuit-and-evasion; the fascination simply becomes more sophisticated. The inherent conflict of chasing and escaping, of the battle between the hunter and the hunted, is the basis for at least half the fictional writing in the world. By their high school years, for example, most students have read the exciting *Collier's Magazine* story "The Most Dangerous Game" by Richard Connell. That 1924 tale of a shipwrecked man (Sangar Rainsford, a celebrated hunter) has the ironic twist of a famous hunter becoming the hunted. At first he finds what he thinks is salvation from drowning at sea as he stumbles ashore onto a nearby jungle-covered island, inhabited by one General Zaroff, who oddly finds himself regularly visited by shipwrecked sailors — probably because he has installed a false lighthouse to lure passing ships onto rocks. Rainsford soon changes his mind about salvation when he learns that the General provides his "visitors" with hunting clothes, food, and a knife, and then turns them loose back into the jungle. After a three-hour head start the General comes after them, armed only with a small caliber automatic pistol — if the General fails to make a kill within three days he promises to return his elusive prey to the mainland. As he explains to Rainsford, however, he has never failed to make the kill. The rest of the story, assigned to generations of high school English students by teachers who clearly admire Connell's tale, is of the suspenseful chase.

The printed story won a prestigious O'Henry Short Story Award, and was made into an exciting movie of the same name in 1932 (Joel McCrea played Rainsford, with his first name changed for some reason from Sangar to *Bob*!, and Fay Wray played a female role not in the original story; this last point is particularly interesting to note since the film was shot between takes, on the same set, of the classic film *King Kong*, which was released the next year in 1933 — starring, of course, Fay Wray). *The Most Dangerous Game* was remade at least two more times: *A Game of Death* (1946) and *Run for the Sun* (1956). There were numerous obvious derivatives of an inferior nature, as well, for example, the 1961 *Bloodlust!*, in which a mad scientist hunts teenagers and displays his catches in glass cases (this movie is notable for having embarrassed one of its stars, Robert Reed, even more than did his later role as Mike Brady, the perfect father on television's *The Brady Bunch*).

Jumping ahead a few decades, most readers of adventure fiction have enjoyed Frederick Forsyth's masterful 1971 novel *The Day of the Jackal*, in which a mysterious assassin-for-hire attempts to kill Charles de Gaulle, the president of France. We all know that de Gaulle was *not* assassinated in real life but, nonetheless, the writing of the killer's chase of his target — and of the intertwined police chase of the assassin — is so well done you think he's going to pull it off anyway. (The book was made into a successful film of the same title in 1973.) The novel was not the first of its type, being a clever reversal of roles from what I think the perfect pursuit-and-evasion fictional work — British writer Geoffrey Household's 1939 novel *Rogue Male* (made into the 1941 film *Man Hunt*). That work is a first-person, page-turning narrative of a world famous English hunter who makes the mistake of being caught sighting a high-powered telescopic rifle onto the distant figure of an unnamed European dictator. (Only a late-1930s reader as dense as a rock could have failed to recognize Hitler.) He is interrogated by his captors, that is, the Gestapo, and then escapes by a fantastic ruse back to England, where his enemies continue to pursue him ruthlessly. The book has a wonderful ending, telling the reader that matters have most definitely *not* ended and the chase is by no means over. This ending was literally a call-to-arms to the world of 1939 to challenge Hitler and defeat him.

A little thought will convince you that Hollywood continues to love a good chase and (perhaps) escape — who, for example, can think of Steve McQueen's *Bullitt* (1968) without instantly visualizing that incredible car chase through the hilly streets of San Francisco? Other successful "chase" films include *Predator* (1987), *The Hunt for Red October* (1990), *The Fugitive* (1993), *Alien* (1979), *Catch Me If You Can* (2002), and the very best such film of all (in my opinion), Cornel Wilde's powerful portrayal of a naked man on the run from headhunters in his brilliant *The Naked Prey* (1961).

A relatively new (since the early 1980s) form of entertainment, computer video games, has embraced the chase-and-escape concept as the basis for some of the most successful of all such games (e.g., *Thief, Call of Duty, Half-Life, Max Payne, CounterStrike, Medal of Honor,* and *Hitman*, all of which have had sequels to their original appearances). These games, called first-person shooters (so named because the screen image is the "world" as seen through the eyes of the protagonist) can be quite violent — typically, that image includes "your" hand holding a weapon for instance, a knife, gun, or club. The sales pitch for one of the *Hitman* games, for example, is "See the world, meet interesting people, kill them" — in *Hitman* you play as an assassin-for-hire who is assigned targets, mostly international terrorists and crime lords, to hunt for "termination with extreme prejudice." These games can be subtle, too, however — the emphasis in the *Thief* games is not on killing at all (although you *do* carry a sword or dagger, and a bow with arrows), but rather on being able as a medieval thief to skillfully skulk about in the shadows and, if at all possible, to *avoid* confrontation while attempting to complete the mission goals. Whether violent or (relatively) benign, the successful first-person-shooter games *are* fun — many are even educational — to experience: playing as the conscript Alexei Ivanovich (a Russian sniper at the Second World War Battle of Stalingrad) in *Call of Duty* can only be described as both ear-shattering and eye-opening, as well as masterfully capturing the horror of what happened there on the rooftops, in the sewers, and on the bombed-out Stalingrad streets in a way that I think equal to the best historical writing (e.g., Antony Beevor's 1998 *Stalingrad, The Fateful Siege: 1942–1943*).

The mathematical puzzles of chases and escapes were recognized quite early in the history of mathematics, for example, with the paradox given by the fifth-century B.C. Greek mathematician Zeno of Elea, of the race between Achilles (Homer's hero in the *Iliad*) and the tortoise. As Smith (1917) says about the origin of all pursuit problems, "It would be difficult to conceive of a problem that would seem more real, since we commonly overtake a friend in walking, or are in turn overtaken. It would therefore seem very certain that this problem is among the ancient ones in what was once looked upon as higher analysis. We have a striking proof that this must be the case in the famous paradox of Achilles and the Tortoise . . ." (Zeno may well have been inspired to think upon the problem of *linear* pursuit — a chase along a straight line — by the even more ancient Aesop fable of the race between the hare and the tortoise. That tale predates Zeno by at least a century; its intent, unlike Zeno's paradox, was not mathematical at all, but rather to serve as an illustration of the *moral lesson* that "perseverance wins the race.") Zeno's paradox is easy to explain.

Achilles, fleet of foot, runs ten times as fast as the tortoise and so, to make their race more interesting, he gives the tortoise a ten-mile head start. Zeno's paradox is the result of a seductive argument that seems to prove that Achilles can never even catch the tortoise, much less pass him to win the race. This is a conclusion that is seen to be incorrect if one simply watches a real race! And that's the paradox, of course — where does Zeno's argument (which *is* pretty convincing to most young students when they first encounter it) go wrong? Zeno's argument is as follows: by the time Achilles runs through the initial ten mile head start, the tortoise will have run through one mile, and by the time Achilles runs through that one mile lead the tortoise will have run through one-tenth of a mile, and by the time Achilles runs through that one-tenth of a mile lead the tortoise will have run through one-hundreth of a mile, etc., etc., etc. This process continues through an *infinity* of ever decreasing leads that the tortoise has over Achilles and so, concludes Zeno, Achilles can never catch the tortoise.

The flaw in Zeno's argument escaped understanding for a very long time. Intellects as profound as those of Plato and Aristotle were befuddled by the problem, with Aristotle (in his *Physics*) correctly calling it a fallacy without being able to explain *why*. Today, twenty-five centuries later, high school students learn that the answer is that an infinity of numbers (the infinity of times it takes Achilles to run through the infinity of leads the tortoise has) does *not* necessarily have to add up to infinity. To see this, let v be the speed of the tortoise (and so $10v$ is the speed of Achilles). Thus, to run through the ten-mile head start takes Achilles a time of $10/10v$. Then, to run through the now one-mile lead of the tortoise takes Achilles $1/10v$. To run through the now one-tenth-of-a-mile lead takes Achilles $1/10/10v$. To run through the now one-hundreth-of-a-mile lead takes Achilles $1/100/10v$. And so on. So, the total time required for Achilles to catch the tortoise is

$$\frac{10}{10v} + \frac{1}{10v} + \frac{1/10}{10v} + \frac{1/100}{10v} + \cdots = \frac{1}{v}\left(1 + \frac{1}{10} + \frac{1}{100} + \frac{1}{1000} + \cdots\right).$$

The expression in the parentheses is a geometric series, easily summed to $10/9$, that is, the total time it takes Achilles to catch the tortoise is the finite time $10/9v$, which of course we see is simply the time to run through the initial head start of ten miles at Achilles' *closing speed* of $10v - v = 9v$.

A different sort of pursuit-and-evasion problem, of great interest today among computer scientists because it requires skill in handling not a differential equation but rather what is called a *data structure*, also has an ancient origin: the Greek mythological story of Theseus and the Minotaur, which appears in the written word as early as in Ovid's *Metamorphoses* (dating from the time of Christ), although the myth itself is even more ancient, by *centuries*. The Minotaur, the result of an unnatural union between the wife of King Minos of Crete and a bull, was a flesh-eating monster with the head of a bull and the body of a man — but Edith Hamilton's classic book *Mythology* contains an illustration of the Minotaur with the head of a man and the body of a bull. This monster was kept confined by Minos to a complicated maze of connected passages called the Labyrinth, and once a year the creature was given a sacrifice offering of seven youths and seven maidens. These unfortunates were simply tossed into the Labyrinth, which was so complicated that they soon became lost and unable to find their way free. Eventually the Minotaur would come upon them in their confused wanderings and they were quickly devoured. The Minotaur's reign of terror was finally ended by Theseus, who, before entering the Labyrinth, tied one end of a ball of string to the entry. Then, after wandering through the maze until he found the Minotaur asleep — which he then killed — Theseus was able to escape by simply following the string back to the entry point. A popular undergraduate computer science programming project assignment is the writing of the code to simulate Theseus' hunt for the Minotaur through a Labyrinth of any desired structure, that is, one with arbitrarily complex connectivity.

All of the above discussion was to make the case that the subject of this book resonates with a fundamental psychological aspect of human nature. But let's now leave mere fictional games of chases and escapes, and turn our attention to the really good stuff; let's take a long look at the *mathematics* of pursuit-and-evasion. The intellectual demands will be significant; as Davis (1962) wrote, pursuit problems with "their curious difficulties have intrigued the fancy and strained the ingenuity of modern mathematicians." You'll agree by the time you finish, but I also think you'll see it will have been well worth the effort.

Chapter 1

The Classic
Pursuit Problem

1.1 Pierre Bouguer's Pirate Ship Analysis

Modern mathematical pursuit analysis is generally assumed to begin with a problem posed and solved by the French mathematician and hydrographer Pierre Bouguer (1698–1758) in 1732. This general assumption is not quite correct, as I'll soon elaborate, but Bouguer's problem is today nevertheless taken as the starting point of pursuit analysis in all modern textbooks, and I'll do the same here. In his paper, read before the French Academy on January 16, 1732, and published in the Academy's *Mémoires de l'Académie Royale des Sciences* in 1735, Bouguer treated the case of a pirate ship pursuing a fleeing merchant vessel, as illustrated in figure 1.1.1. The pirate ship and the merchant vessel are taken to be at $(0,0)$ and $(x_0, 0)$ at time $t = 0$, respectively, the instant the pursuit begins, with the merchant vessel traveling at constant speed V_m along the vertical line $x = x_0$. The pirate ship travels at constant speed V_p along a curved path such that *it is always moving directly toward the merchant,* that is, the velocity vector of the pirate ship points directly at the merchant vessel at every instant of time. Bouguer's problem was to determine the equation $y = y(x)$ of the curved path which he called the *courbe* (or *ligne*) *de poursuite,* the *curve*

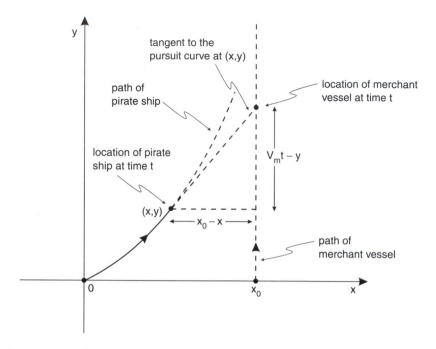

Figure 1.1.1 The geometry of Bouguer's pursuit problem

or *line of pursuit*. As observed in Puckette (1953), the pursuit curve is occasionally called the *courbe de chien* in France and the *verfolgungskurve* or *hundekurven* in Germany, from its early association with the path taken by a dog in following its master. (In Italy it was the *curva di caccia*.) And surely it was noticed from the earliest days of falconry (in Asia, perhaps as long ago as 2000 B.C.) that in its attack the falcon always flies directly at the instantaneous location of its prey; this is the very definition of what is now called *pure pursuit*.

To find the curve of pursuit for Bouguer's problem, start by calling the location of the pirate ship, at arbitrary time $t \geq 0$, the point (x, y). At time t the merchant vessel has sailed to the point $(x_0, V_m t)$ and so, as shown in figure 1.1.1, the slope of the tangent line to the pursuit curve (the value of dy/dx at (x, y)) is given by

(1.1.1)
$$\frac{dy}{dx} = \frac{V_m t - y}{x_0 - x} = \frac{y - V_m t}{x - x_0}.$$

We also know that, whatever the shape of the pursuit curve, the pirate ship has sailed along it at time t by a distance of $V_p t$. From calculus we know that this arc-length is also given by the expression on the right below, and so

$$(1.1.2) \qquad V_p t = \int_0^x \sqrt{1 + \left(\frac{dy}{dz}\right)^2}\, dz,$$

where z, of course, is simply a dummy variable of integration. Solving $(1.1.1)$ and $(1.1.2)$ each for t, we can write

$$\frac{1}{V_p} \int_0^x \sqrt{1 + \left(\frac{dy}{dz}\right)^2}\, dz = \frac{y}{V_m} - \frac{(x - x_0)}{V_m} \cdot \frac{dy}{dx},$$

which, if we let $dy/dx = p(x)$, becomes

$$(1.1.3) \qquad \frac{1}{V_p} \int_0^x \sqrt{1 + p^2(z)}\, dz = \frac{y}{V_m} - \frac{(x - x_0)}{V_m} \cdot p(x).$$

Differentiating $(1.1.3)$ with respect to x (using Leibniz's formula[1] to differentiate an integral), we arrive at

$$\frac{1}{V_p} \sqrt{1 + p^2(x)} = \frac{1}{V_m} \cdot \frac{dy}{dx} - \frac{(x - x_0)}{V_m} \cdot \frac{dp}{dx} - \frac{1}{V_m} p(x)$$

or, simplifying,

$$(1.1.4) \qquad (x - x_0)\frac{dp}{dx} = -\frac{V_m}{V_p} \sqrt{1 + p^2(x)} = -n\sqrt{1 + p^2(x)},$$

where $n = V_m / V_p$. (Ordinarily we'll have $n < 1$, the pirate ship sailing faster than the merchant. For $n > 1$ the problem is without interest as then the pirate ship is slower than the merchant and the concept of "pursuit" is meaningless. The $n = 1$ case, however, *does* offer us a curious mathematical problem with special interest that I'll go into later.) Separating variables,

$$(1.1.5) \qquad \frac{dp}{\sqrt{1 + p^2}} = -\frac{n \, dx}{(x - x_0)} = \frac{n \, dx}{x_0 - x}$$

and, integrating (1.1.5) indefinitely (using a good table of integrals), we have (with C as the constant of indefinite integration)

$$(1.1.6) \qquad \ln\left(p + \sqrt{1 + p^2}\right) + C = -n \ln(x_0 - x).$$

From figure 1.1.1 we see at $t = 0$ that $p = dy/dx = 0$ when $x = 0$ because at that instant both ships are on the x-axis (the fact that $dy/dx|_{t=0} = 0$ also follows *mathematically* from (1.1.1) since $y(t = 0) = 0$). Inserting these initial conditions into (1.1.6), it follows that $C = -n \ln(x_0)$ and so (1.1.6) becomes

$$\ln\left(p + \sqrt{1 + p^2}\right) - n \ln(x_0) = -n \ln(x_0 - x),$$

which, after a few steps of algebra, reduces to

$$\ln\left[\left(p + \sqrt{1 + p^2}\right)\left(1 - \frac{x}{x_0}\right)^n\right] = 0,$$

which tells us that

$$(1.1.7) \qquad \left(p + \sqrt{1 + p^2}\right)\left(1 - \frac{x}{x_0}\right)^n = 1.$$

Thus,

$$(1.1.8) \qquad p + \sqrt{1 + p^2} = \frac{1}{(1 - x/x_0)^n} = q,$$

where q has been introduced to keep the next few algebraic steps easy to follow. Solving (1.1.8) for p, we have

$$\sqrt{1+p^2} = q - p,$$

$$1 + p^2 = (q - p)^2 = q^2 - 2qp + p^2,$$

$$p = \frac{q^2-1}{2q} = \frac{1}{2}\left[q - \frac{1}{q}\right].$$

Thus, replacing q with its equivalent (from (1.1.8)) gives

(1.1.9) $$p(x) = \frac{dy}{dx} = \frac{1}{2}\left[\left(1 - \frac{x}{x_0}\right)^{-n} - \left(1 - \frac{x}{x_0}\right)^{n}\right], \quad n = \frac{V_m}{V_p}.$$

We can solve (1.1.9) for $y(x)$ by simple integration, writing C once more as the constant of indefinite integration,

$$y(x) + C = \frac{1}{2}\int \frac{dx}{(1 - x/x_0)^n} - \frac{1}{2}\int \left(1 - \frac{x}{x_0}\right)^{n} dx.$$

In both integrals change variable to $u = 1 - x/x_0$ (so $dx = -x_0 du$) to get

(1.1.10) $$y(x) + C = \frac{1}{2}\int \frac{-x_0 du}{u^n} - \frac{1}{2}\int -x_0 u^n du,$$

which immediately integrates to

$$y(x) + C = -\frac{1}{2}x_0\frac{u^{-n+1}}{-n+1} + \frac{1}{2}x_0\frac{u^{n+1}}{n+1}$$

$$= \frac{1}{2}x_0\left[\frac{u \cdot u^n}{1+n} - \frac{u \cdot u^{-n}}{1-n}\right] = \frac{1}{2}x_0 u\left[\frac{u^n}{1+n} - \frac{u^{-n}}{1-n}\right].$$

That is,

$$y(x) + C = \frac{1}{2}x_0\left(1 - \frac{x}{x_0}\right)\left[\frac{(1-x/x_0)^n}{1+n} - \frac{(1-x/x_0)^{-n}}{1-n}\right]$$

or

(1.1.11) $$y(x) + C = \frac{1}{2}(x_0 - x) \left[\frac{(1 - x/x_0)^n}{1 + n} - \frac{(1 - x/x_0)^{-n}}{1 - n} \right].$$

Since $y(x = 0) = 0$, then

$$C = \frac{1}{2}x_0 \left[\frac{1}{1 + n} - \frac{1}{1 - n} \right] = -\frac{n}{1 - n^2}x_0$$

and so inserting this result into (1.1.11) gives us our answer, the pursuit curve equation $y = y(x)$:

(1.1.12) $$y(x) = \frac{n}{1 - n^2}x_0 + \frac{1}{2}(x_0 - x)$$
$$\times \left[\frac{(1 - x/x_0)^n}{1 + n} - \frac{(1 - x/x_0)^{-n}}{1 - n} \right], \quad n = \frac{V_m}{V_p}.$$

"Capture" occurs when $x = x_0$ (the pirate ship pursuit curve intersects the merchant's course), which says capture occurs at the point $(x_0, n/(1 - n^2)x_0)$. (This makes physical sense only if $n < 1$, of course, the case of the pirate ship being faster than the merchant.) For example, if the pirate ship sails twice as fast as the merchant then $n = \frac{1}{2}$ and capture occurs at the point $(x_0, \frac{2}{3}x_0)$, while if the pirate ship sails only one-third faster than the merchant (i.e., $V_p = \frac{4}{3}V_m$) then $n = \frac{3}{4}$ and capture occurs at the point $(x_0, \frac{12}{7}x_0)$. As n approaches one, that is, as the sailing speeds of the pirate ship and the merchant vessel become equal, it is clear that the capture point moves ever farther up the $x = x_0$ line and, in the limit $n = 1$, the capture point is at infinity (which is the physically obvious statement that capture does *not* occur). Figure 1.1.2 shows the pursuit curve up to the capture point for the case of $x_0 = 1$ and $n = \frac{3}{4}$.

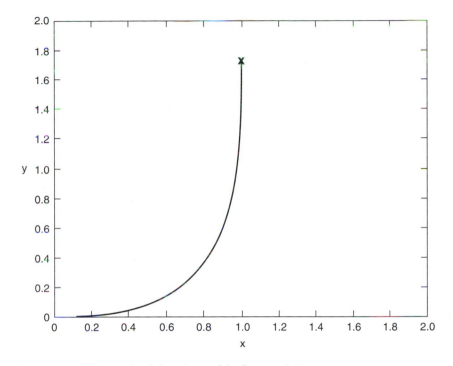

Figure 1.1.2 The path of the pirate ship for $n = 3/4$

The analytical expression of (1.1.12) fails to make sense for the case of $n = 1 (V_m = V_p)$, of course, because then we have a division by zero problem. To see what the correct analytical form of the pursuit curve is for $n = 1$, return to (1.1.9), to just *before* we integrated dy/dx. Then

$$\frac{dy}{dx} = \frac{1}{2}\left[\left(1 - \frac{x}{x_0}\right)^{-1} - \left(1 - \frac{x}{x_0}\right)\right] = \frac{1}{2}\left[\frac{1}{(1 - x/x_0)} - \left(1 - \frac{x}{x_0}\right)\right]$$

(1.1.13)

and so

$$y(x) + C = \frac{1}{2}\left[\int \frac{dx}{1 - x/x_0} - \int \left(1 - \frac{x}{x_0}\right)dx\right].$$

As before, change variables in both integrals to $u = 1 - x/x_0$ (and so $dx = -x_0 \, du$) to get

$$y(x) + C = \frac{1}{2} \int \frac{-x_0}{u} du - \frac{1}{2} \int u(-x_0) du$$

$$= -\frac{1}{2} x_0 \ln u + \frac{1}{2} x_0 \cdot \frac{1}{2} u^2$$

$$= \frac{1}{2} x_0 \left[\frac{1}{2} \left(1 - \frac{x}{x_0} \right)^2 - \ln \left(1 - \frac{x}{x_0} \right) \right].$$

Since $y(x = 0) = 0$, then $C = \frac{1}{4} x_0$, and so for $n = 1 (V_p = V_m)$ the equation of the pursuit curve is

$$(1.1.14) \qquad \boxed{ y(x) = \frac{1}{2} x_0 \left[\frac{1}{2} \left(1 - \frac{x}{x_0} \right)^2 - \ln \left(1 - \frac{x}{x_0} \right) \right] - \frac{1}{4} x_0. }$$

When Bouguer's problem was included in the 1859 book *Treatise on Differential Equations* by the famous British mathematician George Boole (1815–1864), the pursuit curve for the $n = 1$ case was declared to be a parabola. The result in (1.1.14) shows Boole's result to be incorrect, but what is really surprising is that he missed the *obvious* fact that a parabola *must* be wrong — as observed in Burton and Eliezer (2000), whatever the pursuit curve is (for any value of n) it certainly must be asymptotic to the line $x = x_0$, a property no parabola could possibly have.

Two Challenge Calculations

For $n < 1$, what is the total distance traveled by the pirate ship until its capture of the merchant vessel? (Don't make things harder than they are — this is actually an *easy* problem!) Also, as discussed earlier, capture does *not* occur in the $n = 1$ case; rather, after a "long" time, the pirate ship will have sailed into a position directly behind the merchant and will simply chase, endlessly, after the merchant while remaining a *constant distance* behind it. It is an interesting mathematical problem to calculate the value of this so-called *tail chase lag distance*. If you run into trouble (or simply want to check your answers), then take a look at appendix A.

Now, here's an application example of the previous discussions. In December of each year the Mathematical Association of America sponsors the William Lowell Putnam Mathematical Competition on college campuses all across North America. It is a *very* tough test — with twelve problems, each worth ten points, the median score is typically just *one* point — although in the history of the exam there have been three *perfect* scores, too! The 1959 Putnam Competition included the following problem (which originally appeared in the November 1932 issue of *American Mathematical Monthly*):

> A sparrow, flying horizontal in a straight line, is 50 feet directly below an eagle and 100 feet directly above a hawk. Both hawk and eagle fly directly toward the sparrow, reaching it simultaneously. The hawk flies twice as fast as the sparrow. At what rate does the eagle fly?

This is a nice twist on Bouguer's problem, for which we've already done the heavy lifting.

The statements that the eagle is above the sparrow, and that the hawk is below, are red herrings and have nothing to do with the *mathematics* of the problem. We can, with no loss in the spirit of the problem, take the initial location of the eagle as $(0, e)$ and of the hawk as $(0, h)$, where $e = 50$ and $h = 100$. In our solution to Bouguer's problem, the initial separation between pursuer and pursued was x_0, and so e and h each play the role of x_0. We know from our earlier analysis that capture will occur after the pursued has traveled distance of $(n/1 - n^2)x_0$, where $n = $ speed of pursued/speed of pursuer. For the hawk we have $n = \frac{1}{2}$, and for the eagle let's say it flies k times as fast as the sparrow (and so $n = 1/k$ for the eagle). Now, since both pursuers "capture" the sparrow at the same instant (the same point) we have

$$\frac{1/2}{1-(1/2)^2}h = \frac{1/k}{1-(1/k)^2}e.$$

That is,

$$\frac{1/2}{3/4}100 = \frac{k}{k^2-1}50 = \frac{200}{3},$$

or,

$$4k^2 - 4 = 3k,$$

$$4k^2 - 3k - 4 = 0,$$

$$k = \frac{3 \pm \sqrt{9+64}}{8} = \frac{3 \pm \sqrt{73}}{8},$$

or, as $k > 0$ — actually, we know k must be between one and two (the eagle must fly faster than the sparrow to capture it but slower than the hawk since the eagle starts closer to the sparrow than does the hawk) — we use the plus sign to conclude that the eagle flies 1.443 times as fast as the sparrow.

When Bouguer's problem was published in 1735 it was immediately followed by a generalized analysis of ship pursuit — obviously inspired by Bouguer's work — written by the French mathematician and astronomer Pierre-Louis de Maupertuis (1698–1759). As Bernhart (1954) put it, "Maupertuis finds a short route to Bouguer's equations, but does not study either the differential equation [of the pursuit] or the path which it represents," and so it is *Bouguer* who rightfully deserves the credit for originating mathematical pursuit studies (but see section 1.4). See also Bernhart (1959b). It didn't take long for Bouguer's problem to get into textbooks: Ball (1921) writes "The general problem of the form of the [pursuit] curve, the pursued travelling in a straight line, is discussed in T. Simpson's [the English mathematician Thomas Simpson (1710–1761)] *New Treatise of Fluxions*, London, 1737, p.170," just five years after Bouguer first posed and solved the problem.

As a final comment on Bouguer's problem, a different generalized form of it was solved in Coleman (1991), in which the merchant vessel's straight sailing path is inclined from the vertical by angle α. The solution presented in this section of the book is of course for $\alpha = 0$, while $\alpha = \pi/2$ radians would represent the merchant sailing directly *away* from the pirate ship (and $\alpha = -\pi/2$ radians would represent the merchant sailing directly *toward* the pirate ship). In both of these extreme cases the pursuit curve is, by inspection, simply $x = 0$ (the x-axis), but for $\alpha \neq \pm\pi/2$ or 0 the pursuit curve is pretty complicated, and its derivation is an exercise in nontrivial (i.e., tedious) algebraic manipulation. Coleman's paper prompted the interesting reply by Eliezer and Barton (1992). See also Luther (1941).

1.2 A Modern Twist on Bouguer

Pirates chasing after fleeing merchant ships was okay in Bouguer's day, but for modern students a more up-to-date, relevant context is desirable. An extreme example of this can be found in Hoenselaers (1995), in which a dog executes pure pursuit against a *relativistically* fast rabbit, that is, the dog always runs toward where he *sees* the rabbit — who of course isn't *there* because the rabbit's speed is a big fraction of the speed of light! This is a mathematically significant problem, but, I fear, most students would find it even more fanciful than pirates chasing after merchants. For our modern version of Bouguer's problem, therefore, I'll be a bit more conservative (i.e., realistic).

But not as conservative as was William Holding Echols (1859–1934), a professor of mathematics at the University of Virginia. In his 1902 *Elementary Textbook on the Differential and Integral Calculus*, he included the following curious problem:

> A hawk can fly v feet per second, a hare can run v' feet per second. The hawk, when a feet vertically above the hare, gives chase and catches the hare when the hare has run b feet. Find the length of the curve of pursuit.

This problem is actually *trivial* (at least it is for a book with Echols's title). After all, the hare runs b feet in b/v' seconds, which is the duration of the hawk's flight (at a speed of v feet per second) and so, with *minimal* calculation effort, we see that the length of the hawk's flight is simply bv/v' feet. This is a little *too* elementary for us, here! For the modern student we need a compromise between

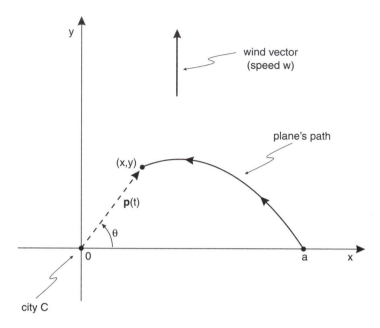

Figure 1.2.1 The geometry of the wind-blown plane problem

Hoenselaer's super-fast rabbit and Echols's routine problem of simple proportions.

So, here's a "modernized" version of Bouguer's problem that I have found in a number of contemporary textbooks:[2]

> A pilot always keeps the nose of his plane pointed toward a city C due west of his starting point at $(a, 0)$. If the plane's speed is v mi/h, and a wind is blowing from the south at the rate of w mi/h, show that the equation of the plane's path is [I'll save the answer until the end of the derivation].

The pilot isn't really *pursuing* anything, of course, unless you consider this a problem of "pursuit (with wind interferrence) of a stationary target," but the spirit of this problem is pure Bouguer. The author I quoted above marked this as one of the more challenging problems in the problem set in which it appears and, as you'll see, it does require a new trick or two to solve.

In the notation of figure 1.2.1, at an arbitrary time $t \geq 0$, the plane's location is the point $(x(t), y(t))$. Writing $\mathbf{u}_x$ and $\mathbf{u}_y$ as the unit vectors in the x and y directions (which are *not* functions of time), respectively, then we can write the *position vector* of the plane as

$$\mathbf{p}(t) = x(t)\mathbf{u}_x + y(t)\mathbf{u}_y,$$

and so the plane's *velocity vector* is

$$\frac{d}{dt}\mathbf{p}(t) = \frac{dx}{dt}\mathbf{u}_x + \frac{dy}{dt}\mathbf{u}_y.$$

Also, the plane's body axis (nose-to-tail) is always along the direction of $\mathbf{p}(t)$, at angle θ, toward C, where

$$\tan(\theta) = \frac{y}{x}.$$

The wind, blowing only along the y-axis, contributes nothing to the $\mathbf{u}_x$ component of the plane's velocity vector, that is, dx/dt is due *only* to the x-component of v;

(1.2.1)
$$\frac{dx}{dt} = -v\cos(\theta) = -\frac{vx}{\sqrt{x^2+y^2}},$$

where the minus sign is explicitly included because as the plane flies toward C the value of x *decreases* with increasing t. The $\mathbf{u}_y$ component of the plane's velocity vector, on the other hand, *is* influenced by the wind, of course, as well as by the y-component of v,

(1.2.2)
$$\frac{dy}{dt} = w - v\sin(\theta) = w - \frac{vy}{\sqrt{x^2+y^2}} = \frac{w\sqrt{x^2+y^2} - vy}{\sqrt{x^2+y^2}}.$$

Dividing (1.2.2) by (1.2.1), we eliminate explicit time and arrive at

(1.2.3)
$$\frac{dy}{dx} = \frac{vy - w\sqrt{x^2+y^2}}{vx},$$

which at first glance certainly doesn't look very encouraging. However, this differential equation yields to the following clever trick.

We introduce a new variable z such that $y = zx$. Then (1.2.3) becomes

$$\frac{dy}{dx} = z + x\frac{dz}{dx} = \frac{vzx - w\sqrt{x^2 + z^2 x^2}}{vx} = z - \frac{w}{v}\sqrt{1 + z^2}$$

or, defining the *constant* $n = w/v$,

(1.2.4)
$$x\frac{dz}{dx} = -n\sqrt{1 + z^2}.$$

Now *this* is a differential equation that offers us hope, as it is separable, that is,

(1.2.5)
$$\frac{dz}{\sqrt{1 + z^2}} = -n\frac{dx}{x}.$$

(Notice the similarity of (1.2.5) with (1.1.5).) Integrating indefinitely, with C as the usual constant of indefinite integration,

$$\ln[(z + \sqrt{1 + z^2})] + C = -n\ln(x).$$

Since $y = 0$ when $x = a$, which means $z = y/x = 0$ when $x = a$, then we have $C = -n\ln(a)$, and so

$$\ln[(z + \sqrt{1 + z^2})] = n\ln(a) - n\ln(x) = n\ln\left(\frac{a}{x}\right) = \ln\left(\frac{a}{x}\right)^n,$$

or,

(1.2.6)
$$z + \sqrt{1 + z^2} = \left(\frac{a}{x}\right)^n.$$

Defining $q = (a/x)^n$, (1.2.6) becomes (look again at how we went from (1.1.8) to (1.1.9))

$$\sqrt{1 + z^2} = q - z,$$

$$1 + z^2 = q^2 - 2qz + z^2,$$

$$z = \frac{q^2 - 1}{2q} = \frac{1}{2}\left[q - \frac{1}{q}\right].$$

Thus, replacing q with its definition,

(1.2.7) $$z = \frac{1}{2}\left[\left(\frac{a}{x}\right)^n - \left(\frac{a}{x}\right)^{-n}\right] = \frac{1}{2}\left[\left(\frac{x}{a}\right)^{-n} - \left(\frac{x}{a}\right)^n\right].$$

Since $y = zx$, then

$$y = \frac{1}{2}\left[\frac{x^{-n+1}}{a^{-n}} - \frac{x^{n+1}}{a^n}\right] = \frac{1}{2}\left[\frac{x^{-n+1}}{a^{-n+1}/a} - \frac{x^{n+1}}{a^{n+1}/a}\right]$$

or, at last, we have the equation of the wind-blown plane's path:

(1.2.8) $$y(x) = \frac{a}{2}\left[\left(\frac{x}{a}\right)^{-n+1} - \left(\frac{x}{a}\right)^{n+1}\right], \quad n = \frac{w}{v}.$$

When $n = 0$ — that is, when there is no wind — (1.2.8) collapses to the physically obvious $y(x) = 0$, which simply says that the plane moves directly to city C while always remaining on the x-axis. And when $n = 1$ (when the wind speed equals the plane's speed in still air), the plane's path is the *parabola*

$$y(x) = \frac{a}{2}\left[1 - \left(\frac{x}{a}\right)^2\right].$$

In this case when $x = 0$ we see that $y(0) = a/2$, that is, the plane does *not* reach city C. This probably makes intuitive sense, too, but it is interesting to see that the miss distance is so large. What happens, physically, in the $n = 1$ case, is that the plane arrives at the y-axis with a zero velocity component in the x-direction (notice that the plane's body axis has rotated through an angle of $\theta = 90°$, and then recall (1.2.1)), and so there the plane remains, *motionless* at the point $(0, a/2)$, as it flies directly into the wind with the two equal magnitude but oppositely directed velocity vectors precisely canceling each other. Figure 1.2.2 shows the plane's path for $a = 1$ for several different values of n, and it is clear that for $n < 1$ (≥ 1) the plane reaches (does *not* reach) city C.

Now, as three challenge problems for you to try your hand at, calculate the total flight time of the wind-blown plane for $n < 1$, the total distance flown for the case of n "just less" than one, and finally,

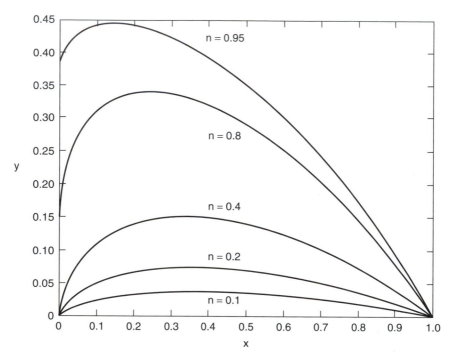

Figure 1.2.2 Some paths of the wind-blown plane

for the third challenge, consider the following clever twist on the wind-blown plane problem. This twist was originally proposed as a challenge question in a 1934 issue of the *American Mathematical Monthly*, and a solution (using a far different approach from that used here) wasn't published until 1936:

> A dog directly opposite his master on the banks of a stream, flowing with uniform speed, swims at a still-water speed of two miles per hour heading directly towards his master at all times. The man notes that the dog does not stop drifting down stream until he is two-thirds across measured perpendicularly to the banks, and that it takes five minutes longer to make the trip than if the water had been still. How wide is the stream? Also, how fast is the stream?

If you get stuck (or just want to check your answers), then take a look at appendix B.[3]

1.3 Before Bouguer: The Tractrix

Forty years before Bouguer's analysis a different sort of pursuit curve had already been discovered, and in the 1700s it was for a time actually called *the* pursuit curve. As you'll see, it would be better to call this curve the *following* curve or (in modern police lingo) the *tailing* curve. Its origin is usually given as the result of an incident in Paris in the late seventeenth century. In a 1693 paper the German mathematician Gottfried Leibniz (1646–1748) recalled an event that occurred sometime during the four years (1672–1676) he spent in Paris studying mathematics with the Dutch mathematician Christiaan Huygens (1629–1695):

> The distinguished Parisian physician Claude Perrault [his brother Charles authored the children's classics *Sleeping Beauty, Little Tom Thumb, Cinderella,* and *Puss in Boots*], equally famous for his work in mechanics and in architecture, well known for his edition of Vitruvius, and in his lifetime an important member of the Royal French Academy of Science, proposed this problem to me and to many others before me, readily admitting that he had not been able to solve it.

Perrault's problem is illustrated in figure 1.3.1, where a watch-on-a-chain has been laid out on a table-top with the chain (of length a) pulled taut. The watch is initially on the y-axis, and the other end of the chain is on the origin. If the end on the origin is then pulled along the x-axis, the watch will obviously be dragged along; the question of interest is, what is the equation of the watch's path? This path — whatever it may be — was given the name *tractrix* by Huygens (the path of the pulled end, in this case the x-axis, is called the *directrix* of the tractrix). Leibniz further recalled that Perrault mentioned to him that no mathematician had been able to solve this problem, not even one from Toulouse (which was Perrault's sly way of saying that the problem had stumped even the great French mathematician Pierre Fermat (1601–1665)). It is interesting to note, however, that Cady (1965) points out Newton had actually solved the tractrix problem *decades* earlier (by 1676).

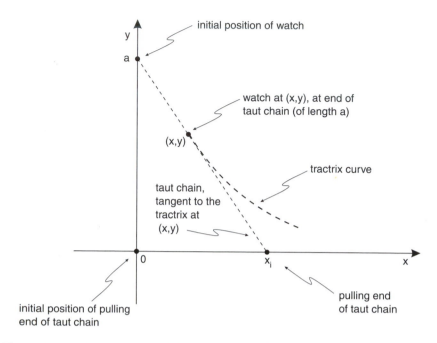

Figure 1.3.1 The geometry of the tractrix problem

If (x, y) is the location of the watch at some arbitrary time $t \geq 0$, then it is clear that the taut chain is *tangent* to the tractrix at (x, y). This crucial observation allows us to calculate where the pulling end of the taut chain is (always on the x-axis, remember), as follows. The slope of the tangent line is dy/dx and so, from analytic geometry, we have the equation of the tangent line as

(1.3.1) $$y = x\frac{dy}{dx} + b,$$

where b is some constant. Let x_i be the value of x where the pulling end of the chain is located; by definition $y = 0$ there. So,

$$b = -x_i \frac{dy}{dx},$$

and therefore the equation of the tangent line that intersects the x-axis at $x = x_i$ is

(1.3.2)
$$y = x\frac{dy}{dx} - x_i\frac{dy}{dx} = (x - x_i)\frac{dy}{dx}.$$

From the Pythagorean theorem we then have

$$(x - x_i)^2 + y^2 = a^2,$$

or, using (1.3.2) to solve for $(x - x_i)$, we have

$$\frac{y^2}{(dy/dx)^2} + y^2 = a^2,$$

or

(1.3.3)
$$\left[\frac{y}{dy/dx}\right]^2 = a^2 - y^2.$$

Taking the positive square root of both sides of (1.3.3), and noting that dy/dx is negative (look at figure 1.3.1 again), we arrive at

(1.3.4)
$$-\frac{y}{dy/dx} = \sqrt{a^2 - y^2},$$

a differential equation in which we can separate the variables. That is,

(1.3.5)
$$dx + \frac{\sqrt{a^2 - y^2}}{y}dy = 0.$$

Integrating indefinitely (with C as the arbitrary constant), we have

$$x + \sqrt{a^2 - y^2} - a \ln\left(\frac{a + \sqrt{a^2 - y^2}}{y}\right) = C.$$

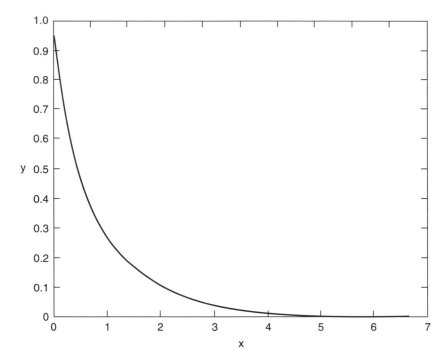

Figure 1.3.2 A tractrix

Since $y(x = 0) = a$, we have $C = 0$ and so the equation of the watch's path as it is being dragged is

(1.3.6)
$$x = a \ln \left(\frac{a + \sqrt{a^2 - y^2}}{y} \right) - \sqrt{a^2 - y^2}.$$

Figure 1.3.2 shows the tractrix of (1.3.6) for the case of $a = 1$.

After reading through this derivation, a modern student might well wonder what the big fuss was all about; "what's so hard about *that?*" is probably what many (you?) are thinking. Well, *today* it isn't hard at all — the hardest part was doing the integration of (1.3.5) to arrive at (1.3.6), which wasn't any harder for me than looking up the answer in a table of integrals. *But* in Perrault's day only a very small number of integrals were known. The tractrix problem *was* hard, *then*, because the integral I looked up so easily in a table at the start of the twenty-first century hadn't been discovered at the end of the seventeenth century! Finally, it is interesting to contrast the tractrix with Bouguer's pure pursuit curve for the special case of equal speeds for the pirate ship and the merchant vessel. The two curves seemingly have a common property, as the dragged watch is a constant distance from the pulled end of the chain, and the pirate ship ends up a constant distance behind the fleeing mechant vessel. The expressions of (1.1.14) and (1.3.6) are quite different, however; the reason is that for the tractrix the constant lag of the watch is *always* the case, while the constant lag of the pirate ship is an *asymptotic* property that *develops* with the passage of time.

1.4 The Myth of Leonardo da Vinci

Before going any farther with our technical discussions of pursuit curves, let me put one story to rest that you'll find repeated in practically all journal papers and textbooks on the subject, right up to this day. It has been the general practice of authors to open their analyses with the throwaway remark "The problem of pursuit seems to have originated with Leonardo da Vinci." After reading that, or some variant of it, numerous times while gathering material for this book, I finally became curious enough about it to see if I could find anything in Leonardo's notebooks or letters specifically related to the pursuit problem. I looked for a long time, but could find *nothing*; but of course that didn't mean anything more than perhaps I hadn't looked in the right place! I remained suspicious of the Leonardo tale, however, for two reasons: (1) no author ever cited a source for the statement (the newer authors were obviously just repeating the older writings), and (2) how *could* Leonardo have said anything about pursuit, beyond simple prose (and he certainly would *not* have been the first to do that) since he lived (1452–1519) *long* before any of the necessary mathematics had even been invented?

I continued to be puzzled by the Leonardo attributions until one day I came across Morley (1921). Frank Vigor Morley (1899–1985), a son of the English-born American mathematician Frank Morley (who will appear in the next chapter), had a great interest in the history of mathematics, and his 1921 paper at last explained to me the origin of the Leonardo story. As he wrote, concerning the problem of pursuit,

> An attempt has been made to make Leonardo da Vinci responsible, among his other wealth of contributions, for the statement of the problem. But although it is quite possible to read into Leonardo's passage the essence of the question, it is perhaps doubtful that he ever had a conscious formulation. And of necessity, careful consideration of the problem had to wait until the methods of the calculus were known.

As a historical scholar, Morley gave a source for what he wrote: an essay by the French army officer and mathematician Henri Brocard (1845–1922). That essay (it appears in an 1880 issue of *Nouvelle correspondance mathématique*) doesn't, as far as I read it, support the claim for Leonardo, as all Brocard does is cite an even earlier (1878) source for *his* claim! A quotation from that source fails, in my opinion, to establish any connection at all between Leonardo and pursuit. Unlike Morley, I am not willing to go even so far as saying "it is quite possible to read into Leonardo's passage — Puckette (1953) dates it at 1510 — the essence of the question." I just don't see it.

Lest any reader feel I am being too harsh on Leonardo, let me assure you that far harsher treatment has already been dished out. In his 1940 book *The Development of Mathematics*, the Caltech mathematician and historian Eric Temple Bell wrote (p. 548), "Leonardo da Vinci's published jottings on mathematics are trivial, even puerile, and show no mathematical talent whatever." That's pretty harsh! But, please, don't get me wrong — Leonardo da Vinci was a genius at what he did, but one of the things he *wasn't* was a mathematician. Let's honor him for what he actually did, and not for what someone long ago *thought* he *might* have done. Pierre Bouguer is the father of the original pure pursuit problem, and it is he who should be the inspiration to all who study pursuit problems today.

1.5 Apollonius Pursuit and Ramchundra's Intercept Problem

This final section of the chapter addresses a question that may have already occurred to you — since the merchant vessel being pursued by Bouguer's pirate ship always sails along a straight line, why does the pirate use *pure pursuit* to run down his victim? Why doesn't the pirate ship simply sail along the straight line path that will *intercept* the merchant? Bouguer himself was not oblivious to that possibility. As Puckette (1953) puts it, "[Bouguer] makes it quite clear that the pursuing ship could catch its quarry much more quickly by 'heading it off' than by merely following it (assuming the line of flight remains a straight line)." So, again, why pure pursuit?

There are at least two answers to that question. First, of course, the pure pursuit problem is simply interesting from a *mathematical* point of view. And second, if the merchant vessel deviates from its straight path and starts executing an active evasion plan, then the pirate ship is going to have to recalculate its intercept course continually anyway; a pure pursuit strategy is just *one* way to specify *how* to do repetitive new course calculations. And, in any case, even for the merchant vessel sticking to a straight line escape path, determining the intercept course for the pirate ship is a nontrivial calculation. In the days of submarine warfare in World War II, for example, this was a most practical problem — submarines fired their torpedoes on *intercept* courses at unsuspecting, that is, nonmaneuvering, enemy surface ships. If you watch old Hollywood war films featuring underwater submarine attacks, then it is almost certain you'll hear the weapons officer yell something like "Intercept solution complete, Captain!," and in response the Captain will order "Fire one, fire two!" It's all very dramatic. Today, it isn't such an important problem because, unlike the "dumb" torpedoes of yesteryear, modern torpedoes use what is called "active tracking," that is, they have onboard sensors and computers that continually locate the target no matter how that target moves. Still, the *mathematics* of interception remains elegant. So, just how *is* that torpedo "intercept solution" calculated?

Let's suppose that the torpedo **T** is to intercept an enemy surface ship **E** (as shown in figure 1.5.1), with **E** moving on a straight path and

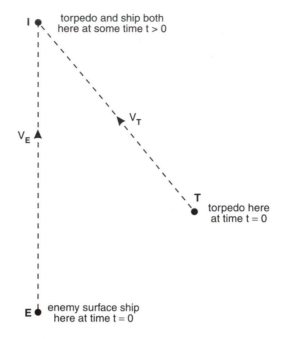

Figure 1.5.1 Pursuit by interception

T moving on a straight path to intercept **E** at point **I**. If we assume that **E** and **T** move with constant speeds V_E and V_T, respectively, then at the intercept point **I** the ratio of the two distances traveled from the instant of the torpedo firing must equal the ratio of the two speeds,

(1.5.1)
$$\frac{IT}{IE} = \frac{V_T}{V_E} = k,$$

where k is a constant ($k > 1$ is the usual case, but the $k < 1$ case will be of interest to us, too, before we are done).

(1.5.1) is the mathematical statement of the physically obvious fact that, for an *interception* to occur, the torpedo and the ship must reach point **I** *simultaneously*. It isn't enough for **E** and **T** to pass through **I** individually — *they must be at* **I** *at the same time*. To find where **I** is, given the locations of **E** and **T** at time $t = 0$, the two speeds V_E and V_T,

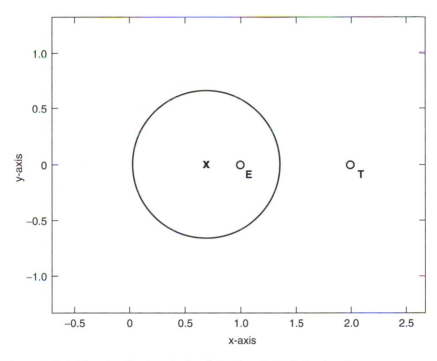

Figure 1.5.2 The Apollonius circle of $(1, 0)$ and $(2, 0)$ for $k = 2$

and the direction of **E**'s motion (the "heading" of **E**), what we must do first is find the set S of all the points in the plane such that (1.5.1) is satisfied. *The* point **I** can be any one of the points (there can be more than one) in **S** that also lie on the path of **E**.

So, what is S? With no loss in generality we can draw a rectangular coordinate system such that **E** and **T** are both, at $t = 0$, on the positive horizontal axis with **T** to the right of **E** (see figure 1.5.2). If we denote the coordinates of **E** and **T** by $(m, 0)$ and $(p, 0)$, respectively, with $p > m$ — I am using m and p here to retain a *small* link with our original discussion of Bouguer's merchant and pirate — and if (x, y) is any point in S, then (1.5.1) becomes

(1.5.2)
$$\frac{\sqrt{(x - p)^2 + y^2}}{\sqrt{(x - m)^2 + y^2}} = k.$$

If you now go through a few algebraic manipulations, then *you* should be able to confirm that (1.5.2) can be written as

$$(1.5.3) \qquad \left[x - \frac{k^2 m - p}{k^2 - 1}\right]^2 + y^2 = \left[\frac{k(p - m)}{1 - k^2}\right]^2.$$

But this is the equation of a *circle*, with its center on the horizontal axis at $((k^2 m - p)/(k^2 - 1), 0)$ and a radius of $k(p - m)/|1 - k^2|$. The set S is a *circle*, called the *Apollonius circle* of the two points **E** and **T** (in their $t = 0$ locations on the horizontal axis), which is named after the third-century B.C. Greek mathematician Apollonius of Perga. Apollonius realized (in his lost work *Plane Loci*) that (1.5.1) is a way to define a circle in a manner different from the usual Euclidean geometry definition (the path traced by a moving point that remains a fixed distance from a given point). The definition in (1.5.1) predates Apollonius, however, being known a century earlier to Aristotle. If $m = 1$, $p = 2$, and $k = 2$, for example, the Apollonius circle is centered on $(\frac{2}{3}, 0)$ with a radius of $\frac{2}{3}$; see figure 1.5.2, where the center of the Apollonius circle is marked with an X and labeled circles indicate the initial locations of the torpedo and the enemy ship. For the submarine to determine where to aim its torpedo (that is, to locate the point **I**), all that remains to do is to see where **E**'s path intersects the Apollonius circle. The intersection point is **I**. For example, you can see from figure 1.5.2 that **I** is, approximately, at $(1, 0.58)$ if **E** has a heading angle of $90°$.

Now, what if $k < 1$, that is, what if the torpedo is *slower* than the surface ship? To be specific, let's now take $k = \frac{1}{2}$, which reduces (1.5.3) (with $m = 1$ and $p = 2$) to

$$\left(x - \frac{7}{3}\right)^2 + y^2 = \left(\frac{2}{3}\right)^2.$$

That is, the Apollonius circle is still of radius $\frac{2}{3}$, but now is centered on $(\frac{7}{3}, 0)$, which means the center of the Apollonius circle is now to the *right* of the initial location of **T**, as shown in figure 1.5.3. You can see that now the torpedo may or may not be able to intercept the enemy ship — it's all a function of the heading angle of the ship. If the heading angle is sufficiently small that the ship's path crosses

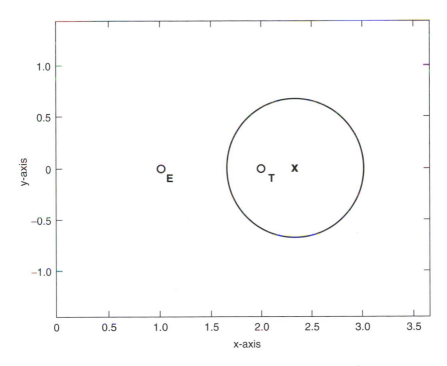

Figure 1.5.3 The Apollonius circle for (1, 0) and (2, 0) for $k = \frac{1}{2}$

the Apollonius circle, then an interception by a slow torpedo of a fast ship *is possible* (in fact, there will generally be *two* possible interception points!), a result that often surprises.

Instead of considering specific values of k, m, and p, it is not at all difficult to be much more general and to derive an astonishingly simple condition that will tell us, for *any* $k < 1$, if a slow torpedo interception is, first, even possible and, if it is, where on the Apollonius circle the submarine should aim its slow torpedo. Equation (1.5.3) tells us that, for $k < 1$, the Apollonius circle for the points $(m, 0)$ and $(p, 0)$, $p > m$, is centered on the point C at $((p - k^2m)/(1 - k^2), 0)$ and has a radius of $k(p - m)/(1 - k^2)$, as illustrated in figure 1.5.4. Now, imagine that the enemy ship's heading is that angle θ that just touches the Apollonius circle at A — if the absolute value of the heading angle is greater than θ then no interception is possible, and if the absolute value of the heading angle is less than θ then the enemy ship's path will cross the Apollonius circle *twice* and so there will be *two* possible

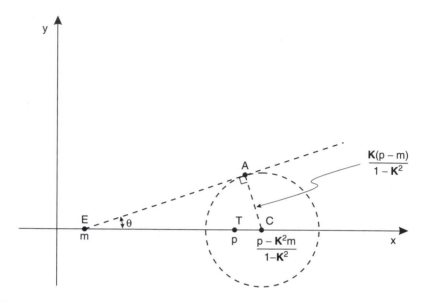

Figure 1.5.4 The general geometry for a slow torpedo interception of a fast surface ship

interception points **I**. We can find a beautifully simple formula for θ, as follows.

The line AC, a radius of the Apollonius circle, is perpendicular to the tangent line EA, and so the triangle ECA is a right triangle. Thus,

(1.5.4)
$$\sin(\theta) = \frac{AC}{EC}.$$

The radius of the circle, as stated before, is

$$AC = \frac{k(p-m)}{1-k^2},$$

while

$$EC = ET + TC = (p-m) + \left(\frac{p-k^2m}{1-k^2} - p \right) = \frac{(p-m)}{1-k^2}.$$

Inserting these expressions for AC and EC into (1.5.4) we arrive at

$$\sin(\theta) = k = \frac{V_T}{V_E},$$

that is,

(1.5.5)
$$\theta = \sin^{-1}\left(\frac{V_T}{V_E}\right).$$

If α is the heading angle of the enemy surface ship, then an interception using a slow torpedo is possible if $-\theta \leq \alpha \leq \theta$, and impossible otherwise.

As a final comment on the intercept problem, there is a little-known mathematical treatment of an interesting variant of it in an odd little book by the obscure nineteenth-century Indian mathematician Yesudas Ramchundra (1821–1880). Published in Calcutta in 1850, and in English (in London, under the sponsorship of the logician Augustus DeMorgan) in 1859, his *Treatise on Problems of Maxima and Minima Solved by Algebra* is an eccentric, incredibly ingenious work that completely avoids the use of the differential calculus when studying questions in extrema. One of the problems discussed by Ramchundra is the intercept problem for a slow pursuer versus a fast target. He wrote,

> Supposing a ship to sail from a given place A, in a given direction AQ, at the same time that a boat from another place B sets out in order (if possible) to come up with her, and supposing the rate at which each vessel progresses to be given [we'll use V_S and V_B for the ship starting at A and the boat starting at B, respectively, with $V_B < V_S$], it is required to find in what direction the latter must proceed, so that if it cannot come up with the former, it may however approach it as near as possible.

This problem statement is illustrated in figure 1.5.5, and while the differential calculus is *our* central mathematical tool in this book, it is still a worthwhile exercise to see how Ramchundra answered the question using nothing but algebra and geometry.

In figure 1.5.5 the points D and F mark the locations of the ship and boat, respectively, when the the two are closest. The figure has been drawn so that BFD is a straight line, but how do we know that is actually the case? Think of it this way (as Ramchundra did): from B imagine a straight line to D (wherever D *is* — we don't know that yet, but it has to be somewhere!). Next, when the boat starting at B

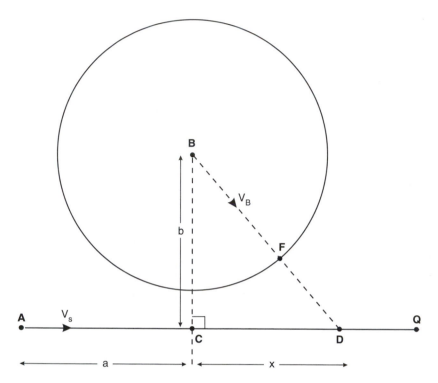

Figure 1.5.5 The geometry of Ramchundra's intercept problem

has reached **F** (wherever *it* is) it has sailed a certain distance, a distance we can use as the radius of a circle centered on **B** (the solid circle). Now, *by definition*, the line **FD** has the least possible length; of all the points on the dashed circle where **F** might be, the one that gives the minimum value to **FD** is the point on the straight line joining **B** and **D**. This is obvious once pointed out, but I'll bet it isn't the first (or even the second) thing most people think of when first confronted with this problem. Ramchundra was a clever fellow, indeed. Finally, to finish the preliminary geometrical set-up of his analysis, imagine a perpendicular line from **B** to **AQ** (defining point **C** on **AQ**), and call the lengths of **AC**, **CD**, and **BC**, a, x, and b, respectively, as shown in figure 1.5.5. Ramchundra then started to write some equations.

First, the length of **BD** is

(1.5.6)
$$BD = \sqrt{BC^2 + CD^2} = \sqrt{b^2 + x^2}.$$

Next, since the time it takes the ship to sail from A to D (at speed V_S) is the time it takes the boat to sail from B to F (at speed V_B), then

$$\frac{AD}{V_S} = \frac{BF}{V_B},$$

or

(1.5.7)
$$BF = \frac{V_B}{V_S}(a+x).$$

At its closest approach to the ship the boat is distance DF from the ship, and so from (1.5.6) and (1.5.7) we have

(1.5.8)
$$DF = BD - BF = \sqrt{b^2+x^2} - \frac{V_B}{V_S}(a+x),$$

where the value of x is whatever it must be for the length of DF to be the smallest possible.

Let's write this minimized value of the length of DF as q; then

$$\sqrt{b^2+x^2} - \frac{V_B}{V_S}x = q + \frac{V_B}{V_S}a.$$

Since $V_B a/V_S$ is a constant, and since q is the smallest that DF can be, then $q + V_B a/V_S$ is the smallest it can be. Let's call this minimized quantity r,

(1.5.9)
$$r = q + \frac{V_B}{V_S}a,$$

and so

$$\sqrt{b^2+x^2} - \frac{V_B}{V_S}x = r,$$

or

$$\sqrt{b^2+x^2} = r + \frac{V_B}{V_S}x,$$

or

$$b^2+x^2 = \left(r + \frac{V_B}{V_S}x\right)^2 = r^2 + 2\frac{V_B}{V_S}xr + \frac{V_B^2}{V_S^2}x^2,$$

which can be rearranged into the quadratric

$$(1.5.10) \qquad x^2 - \frac{2r V_B V_S}{V_S^2 - V_B^2} x - \frac{(r^2 - b^2) V_S^2}{V_S^2 - V_B^2} = 0.$$

Solving for x (and doing a little simplification), *you* should verify that

$$(1.5.11) \qquad x = \frac{r V_B V_S \pm \sqrt{V_S^4 (r^2 - b^2) + V_B^2 V_S^2 b^2}}{V_S^2 - V_B^2}.$$

Now, for the final argument, which is short and sweet: there should be a *unique* value of x that minimizes the length $DF(=q)$, that is, that minimizes r. Thus, the radical in (1.5.11) must vanish, and so

$$V_S^4 (r^2 - b^2) + V_B^2 V_S^2 b^2 = 0$$

or,

$$(1.5.12) \qquad r = \frac{b \sqrt{V_S^2 - V_B^2}}{V_S}.$$

For this value of r (1.5.11) says

$$x = \frac{V_B V_S r}{V_S^2 - V_B^2} = \frac{b V_B \sqrt{V_S^2 - V_B^2}}{V_S^2 - V_B^2},$$

which says the boat is closest to the ship for

$$(1.5.13) \qquad x = \frac{b V_B}{\sqrt{V_S^2 - V_B^2}}, \quad V_B < V_S.$$

This solves the slow boat/fast ship intercept problem: simply drop the perpendicular from point B to point C on the path AQ (this gives us the value of b), then mark off the distance x from C toward Q — that determines D. Then, the sailing path for the slow boat is the straight line joining C to D. That's it! Of course, upon reaching (1.5.8) for DF

as a function of x, a modern student would have simply set

$$\frac{d}{dx}(DF) = 0,$$

which immediately gives the result of (1.5.13). For more on the personal story of Ramchundra, some additional examples from his *Treatise* on solving extrema problems with only algebra and geometry, and for *why* he wanted to avoid anything to do with differentiation, see Raina (1992) and Musès (1998).

I want to emphasize that the presentation of pursuit in terms of pirates chasing after merchants or torpedoes after enemy ships (or missiles after "targets," or Coast Guard boats after rum runners, or lions after men, or monsters after princesses, as will be done in later chapters), is done here purely for dramatic purposes. Mathematicians, actually need no such exciting imagery to formulate interesting pursuit problems. For example, in the late 1870s and 1880s the Italian mathematician Ernesto Cesàro (1859–1906) studied the "pursuit" by the center of gravity of an arc chasing after its moving end (with the other end held fixed). There is no hint of potential violent confrontation implicit in that work — and certainly no hope for capture! — but the mathematics is no less interesting; see Bernhart (1959b). Striking a scholarly compromise between the dramatic and the abstract was the American mathematician Alfred Lotka (1880–1949), in his paper "Families of Curves of Pursuit, and Their Isochrones" (*American Mathematical Monthly*, October 1928). There Lotka showed how to model a so-called "predator-prey" problem as one of pure pursuit. As he wrote, "An organism P_1, of a species which serves as food to an organism P_2, becomes aware of the presence of the latter when at a distance ρ from the same, and immediately flees in a straight line to the nearest cover or refuge, distance D away." P_1 and P_2 might be, for example, a rabbit and a fox, respectively. This particular analysis was just one part of Lotka's pioneering studies in population dynamics, a mathematical topic far removed from explicit violent confrontation.

To end this chapter, here's a modern-day intercept challenge problem for you. Suppose a constant speed, constant altitude attacking missile is spotted by a ground-based defensive site at angle θ above the horizon, heading directly toward the site. Show that if an antimissile missile (flying at the same speed) is immediately launched on a straight line, upward course at angle 2θ, then interception will always occur. Be sure to analyze the problem for *all* physically possible values of θ, $0° < \theta < 90°$ ($0° < 2\theta < 180°$). (Since a launch angle greater than $180°$ is a launch *into the ground*, such a situation makes no physical sense!) You can compare your analysis with mine in appendix C. Assume a flat earth.

Chapter 2

Pursuit of (Mostly) Maneuvering Targets

2.1 Hathaway's Dog-and-Duck Circular Pursuit Problem

In chapter 1 we considered only the case of the pursued (the chased one, the fugitive, the evader, the escaper, the target, etc.) as always moving along a straight path. It didn't take very long after Bouguer's original 1732 statement of the pursuit problem, however, for analysts to begin to consider more complicated evasion paths (what Bouguer called the *ligne de fuite*). The simplest such *flight* or *fleeing* path (in terms of *shape*) would, for most people, be a circular path, and one finds such paths being discussed as early as 1742. These early extensions of the pursuit problem *statement* (but *not* the solution) first appeared in a curious journal, published once a year from 1704 to 1841, called the *Ladies' Diary*. Published in England, the journal was quite popular, with a subscription list of several thousand, and almost surely this was because it devoted a large fraction of its space to intriguing mathematical puzzles. These puzzles were *not* trivial — here's one from the 1783 issue:

$$\text{if } a = \sum_{k=0}^{\infty} \frac{1}{\sqrt{2k+1}}$$

and

$$\text{if } b = \sum_{k=1}^{\infty} \frac{1}{\sqrt{2k}},$$

then what is the exact numerical value of a/b? Try your hand at this before looking at the solution[1] (but be careful, as there is a bit of trickery here!) You can find informative essays on the *Ladies' Diary* in Perl (1979) and Schiebinger (1989).

In the 1748 *Diary*, according to Ball (1921), a man named John Ash stated a circular pursuit in the form of a spider chasing a fly around the edge of a semicircular pane of glass. Ash's problem[2] caught the attention of at least one other, because the *Diary*'s editor later wrote that a reader had indeed

> sent us a true method; but the calculus being so operose, it was not wrought out [which I take to mean that the reader couldn't solve his equations, a fate that would be prove to be that of many analysts to come]. And no method appearing to us yet elegant enough for a place, it will be next year before we have time to catch the solution.

It would take, alas, a lot longer than a year — certainly nothing more appeared in the *Diary* on the circular pursuit problem.

The problem of circular pursuit percolated for another century, until it surfaced again in the mathematical literature in the April 1859 issue of the short-lived *Mathematical Monthly* (it lasted three years). There the editor John Daniel Runkle (1822–1902) — who was later the second president of MIT — observed that Maupertuis's generalized differential equation for Bouguer's problem had never "been integrated even in the simplest case, that in which the given curve [of escape] is a circle," and again in 1877 when Henri Brocard — the man you'll recall from chapter 1 who started the myth of Leonardo da Vinci — challenged readers of *Nouvelle correspondence mathématique* to derive the equation for the pursuit curve in circular pursuit. When no solution was forthcoming, he backed off to asking for just the differential equation of the curve (*that* question is much easier; it *was* answered in 1886, and we'll do it later in this chapter). The problem of *integrating* that (or an equivalent) differential equation continued to be revisited, with no success, however. In 1894, for example, one contributor to the *American Mathematical Monthly* had A pursuing B, with A starting at the center of the circular path around which B is running; the resulting differential equation was so complicated that soon after two attempted solutions were accompanied with the despairing words "this results in a very complicated equation, which has never, so far as I know, been solved," and "all of these equations, as to their integrability, transcend the present limits of mathematical genius."[3]

Yet another quarter-century passed, and then in 1920 the problem of circular pursuit was again proposed in the form it now appears in modern textbooks on differential equations. As A. S. Hathaway wrote in *American Mathematical Monthly* in the January 1920 issue,

A dog at the center of a circular pond makes straight for a duck which is swimming [counterclockwise] along the edge of the pond. If the rate of swimming of the dog is to the rate of swimming of the duck as $n : 1$, determine the equation of the curve of pursuit and the distance the dog swims to catch the duck.

Hathaway was actually not the first to use amusing animal imagery to illustrate circular pursuit — in a 1910 issue of the *Monthly* a reader also had a dog chasing a duck in a circular pond, and in 1902 another reader had a dog running down the surface of a right conical hill toward a fox running around the circular foot of the hill. In the February 1894 issue of *Revue de mathématiques spéciales*, yet another version appeared, which had a jockey trying to catch a horse running around a circular track. It was Hathaway's problem that stuck to become the favorite of textbook authors, however. This is probably due to two reasons: first, the year after his statement of the problem Hathaway published a very complete geometric *analysis* which was included in Archibald and Manning (1921). And second, Hathaway was not just any "reader," but was a well-known name in mathematics

— he was Professor Arthur Stafford Hathaway (1855–1934), on the faculty at Rose Polytechnic Institute in Terre Haute, Indiana, and the author of scholarly works in quaternions and Newtonian dynamics.[4] But still, even Hathaway could not find an analytical expression for the curve of pursuit in circular pursuit. The reason is straightforward — the differential equation for the pursuit curve problem, with the pursued moving on a circle, simply can *not* be expressed in terms of elementary functions. With the aid of a modern computer, of course, this is not even a minor problem when it comes to actually generating plots of the pursuit curve for a variety of conditions.

But, first, even with a computer, we do still need a differential equation before we can apply the power of the machine. To set the problem up mathematically, I'll use the notation of figure 2.1.1. There we have the path of the duck as a circle with radius a, where I'll imagine that the duck starts at time $t = 0$ on the x-axis at the point $(a, 0)$. There is no loss of generality in making this assumption because, no matter where on the circle the duck happens to be at $t = 0$, we'll simply construct our coordinate system so that the duck's position *is* $(a, 0)$. Now, at some time $t > 0$ the duck will have moved through an arc of angle θ (through a *distance* of $a\theta$) and so the dog (who swims n times faster) will have moved through a distance of $s = an\theta$ to reach the point (x, y). Because the dog is executing a pure pursuit, the tangent at (x, y) to the dog's pursuit curve will (by definition) pass through the duck's instantaneous position. As shown in figure 2.1.1, this tangent line makes angle ω with the x-axis, and the distance between the dog and the duck is ρ.

Now, the tangent line through (x, y) has slope $\tan(\omega)$, and so from analytic geometry we can write (with b some constant) the equation of the tangent line as

$$(2.1.1) \qquad y = x \tan(\omega) + b = x \frac{\sin(\omega)}{\cos(\omega)} + b.$$

Since the tangent line passes through the duck's location at $(a \cos(\theta), a \sin(\theta))$, then

$$a \sin(\theta) = a \cos(\theta) \frac{\sin(\omega)}{\cos(\omega)} + b,$$

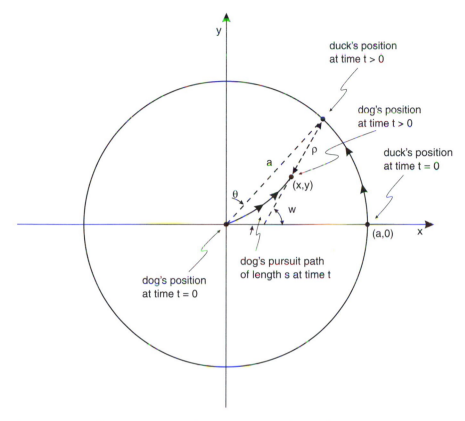

Figure 2.1.1 The geometry of Hathaway's pursuit problem

or

$$b \cos(\omega) = a \sin(\theta) \cos(\omega) - a \cos(\theta) \sin(\omega)$$

$$= a \sin(\theta - \omega) = -a \sin(\omega - \theta).$$

Thus, inserting this result into (2.1.1), we arrive at the equation of the tangent line to the dog's curve of pursuit:

(2.1.2)
$$x \sin(\omega) - y \cos(\omega) = a \sin(\omega - \theta).$$

Now that we have the equation of the tangent to the dog's pursuit curve, our next step will be to determine the equation of the line *normal*

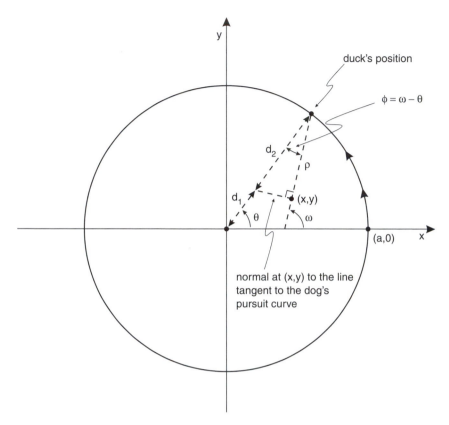

Figure 2.1.2 Finding the normal to the tangent line

to the tangent line that passes through (x, y). Figure 2.1.2 shows the geometry of the calculation. From analytic geometry we know that the slope of the normal will be the negative reciprocal of the tangent line's slope: the slope of the normal line is $-1/\tan(\omega) = -\cos(\omega)/\sin(\omega)$. Thus, the equation of the normal line has the form (with b again some constant)

$$(2.1.3) \qquad y = -x\frac{\cos(\omega)}{\sin(\omega)} + b.$$

As before, to determine b we need to know the coordinates of another point through which the normal line passes. Here's how to find such a point.

The radius from the center of the circular pond to the duck's position is $d_1 + d_2 = a$, where

$$d_2 = \frac{\rho}{\cos(\omega - \theta)}.$$

This defines the triangle with hypotenuse d_1 shown in figure 2.1.2, which allows us to calculate the coordinates of the point at which the normal to the tangent line intersects the radius to the duck's position. We see that

$$d_1 = a - \frac{\rho}{\cos(\omega - \theta)} = \frac{a\cos(\omega - \theta) - \rho}{\cos(\omega - \theta)},$$

and that the x and y values of the intersection point are

$$x = \cos(\theta)\frac{a\cos(\omega - \theta) - \rho}{\cos(\omega - \theta)}, \quad y = \sin(\theta)\frac{a\cos(\omega - \theta) - \rho}{\cos(\omega - \theta)}.$$

Inserting these expressions into (2.1.3) gives us

$$\sin(\theta)\frac{a\cos(\omega - \theta) - \rho}{\cos(\omega - \theta)} = -\cos(\theta)\frac{a\cos(\omega - \theta) - \rho}{\cos(\omega - \theta)} \cdot \frac{\cos(\omega)}{\sin(\omega)} + b.$$

From this we immediately have

$$b = \frac{[a\cos(\omega - \theta) - \rho][\sin(\theta)\sin(\omega) + \cos(\theta)\cos(\omega)]}{\sin(\omega)\cos(\omega - \theta)}$$

or, as $\sin(\theta)\sin(\omega) + \cos(\theta)\cos(\omega) = \cos(\omega - \theta)$, then

$$b = \frac{a\cos(\omega - \theta) - \rho}{\sin(\omega)}.$$

Thus, (2.1.3) becomes

$$y = -x\frac{\cos(\omega)}{\sin(\omega)} + \frac{a\cos(\omega - \theta) - \rho}{\sin(\omega)},$$

and so the equation of the line *normal* to the tangent to the dog's curve of pursuit is

$$(2.1.4) \qquad x\cos(\omega) + y\sin(\omega) = a\cos(\omega - \theta) - \rho.$$

We can complete our analysis for the derivation of the differential equation of the dog's pursuit curve by differentiating (2.1.2) and (2.1.4), each with respect to θ. For (2.1.2) this gives

$$\frac{dx}{d\theta}\sin(\omega) + x\cos(\omega)\frac{d\omega}{d\theta} - \frac{dy}{d\theta}\cos(\omega) + y\sin(\omega)\frac{d\omega}{d\theta},$$

$$= a\cos(\omega - \theta)\left(\frac{d\omega}{d\theta} - 1\right),$$

or

$$\frac{dx}{d\theta}\sin(\omega) - \frac{dy}{d\theta}\cos(\omega) + \frac{d\omega}{d\theta}[\,x\cos(\omega) + y\sin(\omega)]$$

$$= a\cos(\omega - \theta)\left(\frac{d\omega}{d\theta} - 1\right).$$

Since the last factor on the left-hand-side (in square brackets) is $a\cos(\omega - \theta) - \rho$ by (2.1.4), we have

$$\frac{dx}{d\theta}\sin(\omega) - \frac{dy}{d\theta}\cos(\omega) + a\cos(\omega - \theta)\frac{d\omega}{d\theta} - \rho\frac{d\omega}{d\theta}$$

$$= a\cos(\omega - \theta)\frac{d\omega}{d\theta} - a\cos(\omega - \theta),$$

or,

$$(2.1.5) \qquad \frac{dx}{d\theta}\sin(\omega) - \frac{dy}{d\theta}\cos(\omega) - \rho\frac{d\omega}{d\theta} = -a\cos(\omega - \theta).$$

Now, with reference to figure 2.1.3, we see that the variables x and y (which at the dog's position simultaneously satisfy the pursuit curve, the pursuit curve tangent, *and* the pursuit curve normal) are related by

$$\frac{dy}{dx} = \tan(\omega).$$

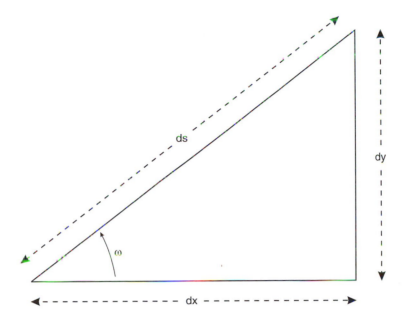

Figure 2.1.3 The "differential triangle" relating ds, dy, and dx

Also, since the dog's travel distance is (from before) given by $s = an\theta$, and since (again, see figure 2.1.3)

$$\frac{dx}{ds} = \cos(\omega),$$

then

$$ds = \frac{dx}{\cos(\omega)} = an\,d\theta$$

and so

(2.1.6) $$\frac{dx}{d\theta} = an\cos(\omega).$$

Also, by the chain rule of calculus,

$$\frac{dy}{d\theta} = \frac{dy}{dx} \cdot \frac{dx}{d\theta} = \tan(\omega)an\cos(\omega),$$

or

(2.1.7) $$\frac{dy}{d\theta} = an\sin(\omega).$$

Inserting (2.1.6) and (2.1.7) into (2.1.5), we have

$$an \cos(\omega) \sin(\omega) - an \sin(\omega) \cos(\omega) - \rho \frac{d\omega}{d\theta} = -a \cos(\omega - \theta),$$

or, at last, we arrive at the differential equation

(2.1.8)
$$\rho \frac{d\omega}{d\theta} = a \cos(\omega - \theta).$$

This isn't a total description of the problem, of course, as we have *two* variables that locate the dog, ω and ρ, that is, we need another differential equation.

We get it by now differentiating (2.1.4):

$$\frac{dx}{d\theta} \cos(\omega) - x \sin(\omega) \frac{d\omega}{d\theta} + \frac{dy}{d\theta} \sin(\omega) + y \cos(\omega) \frac{d\omega}{d\theta}$$
$$= -a \sin(\omega - \theta) \left(\frac{d\omega}{d\theta} - 1 \right) - \frac{d\rho}{d\theta},$$

or

$$\frac{dx}{d\theta} \cos(\omega) + \frac{dy}{d\theta} \sin(\omega) - \frac{d\omega}{d\theta} [x \sin(\omega) - y \cos(\omega)]$$
$$= -a \sin(\omega - \theta) \left(\frac{d\omega}{d\theta} - 1 \right) - \frac{d\rho}{d\theta}.$$

Since the last factor on the left-hand side of this equation (in square brackets) is $a \sin(\omega - \theta)$ by (2.1.2), we have

$$\frac{dx}{d\theta} \cos(\omega) + \frac{dy}{d\theta} \sin(\omega) - a \sin(\omega - \theta) \frac{d\omega}{d\theta}$$
$$= -a \sin(\omega - \theta) \frac{d\omega}{d\theta} + a \sin(\omega - \theta) - \frac{d\rho}{d\theta},$$

or

(2.1.9)
$$\frac{dx}{d\theta} \cos(\omega) + \frac{dy}{d\theta} \sin(\omega) = a \sin(\omega - \theta) - \frac{d\rho}{d\theta}.$$

Inserting the results of (2.1.6) and (2.1.7) into (2.1.9), we get

$$an\cos^2(\omega) + an\sin^2(\omega) = a\sin(\omega - \theta) - \frac{d\rho}{d\theta},$$

or, with a little rearragement, we at last arrive at a second differential equation that links our two dependent variables ω and ρ to the independent variable θ (which increases linearly with time):

(2.1.10)
$$\boxed{\frac{d\rho}{d\theta} = a[\sin(\omega - \theta) - n].}$$

The differential equations of (2.1.8) and (2.1.10) are *coupled nonlinear* differential equations and, alas, are *impossible to solve in closed form.*

This may all seem quite discouraging and, up until the age of the high-speed digital computer, it *was* discouraging! When faced with such a situation, the mathematicians of the 18th, 19th, and first half of the 20th centuries (à la Hathaway) either threw up their hands in defeat or resorted to heroic graphical approximation approachs to effectively integrate their awful equations; for example, see Morley (1921). And totally to ruin the day for those mathematicians of yesteryear, all one had to do was observe that the above pair of boxed differential equations were for a duck traveling on a nice, simple *circle*. Just imagine the equations of horror that would result if the duck moved in only a little more complicated way, for example, along an expanding spiral path, or even along an *arbitrarily* complicated path. Well, today matters are decidedly much nicer. In the next section I'll show you how modern mathematicians handle such problems in a routine way — "all" one needs is a gadget that Professor Hathaway and his colleagues of the 1920s could only dream about (and then only if they read the early science fiction magazines); a 21st-century personal computer that sits on the corner of a small desk (I have a small home office and, hence, a small desk) and performs millions of arithmetic operations per second.

2.2 Computer Solution of Hathaway's Pursuit Problem

Figure 2.2.1 shows the position vectors of the dog and duck at some arbitrary time $t \geq 0$, denoted, respectively by $\mathbf{h}(t)$ and $\mathbf{d}(t)$. (Notice that, since the words *dog* and *duck* start with the same letter, I am now using $\mathbf{h}(t)$ to denote the *hound's* position vector.) As in the previous section, $\boldsymbol{\rho}(t)$ represents the separation between the hound and the duck, that is, from vector algebra we have

$$(2.2.1) \qquad\qquad \mathbf{d}(t) = \mathbf{h}(t) + \boldsymbol{\rho}(t).$$

In Cartesian coordinates, we can write

$$(2.2.2) \qquad\qquad \mathbf{d}(t) = x_d(t) + i y_d(t)$$

and

$$(2.2.3) \qquad\qquad \mathbf{h}(t) = x_h(t) + i y_h(t),$$

where $i = \sqrt{-1}$ (this is not as mysterious as it may seem: i is simply a "tag" that says its coefficient is the y-axis coordinate).

Now, by the definition of pure pursuit, the hound's *velocity* vector always points directly at the duck, that is, $d\mathbf{h}(t)/dt$ is a vector pointing directly at the duck at every instant of time. Of course, we already know of a vector that always points from the hound directly at the duck — $\boldsymbol{\rho}(t) = \mathbf{d}(t) - \mathbf{h}(t)$. Thus, the *unit* magnitude vector pointing from the hound directly at the duck at all times is $\boldsymbol{\rho}(t)$ divided by its magnitude:

$$\text{the unit vector from hound to duck} = \frac{\boldsymbol{\rho}(t)}{|\boldsymbol{\rho}(t)|},$$

and, since $|d\mathbf{h}(t)/dt|$ is the *speed* of the hound, the hound's *velocity* vector is

$$(2.2.4) \qquad\qquad \frac{d\mathbf{h}(t)}{dt} = \left| \frac{d\mathbf{h}(t)}{dt} \right| \cdot \frac{\boldsymbol{\rho}(t)}{|\boldsymbol{\rho}(t)|}.$$

The velocity vector of the duck (which by the statement of the problem is *known*; for example, Hathaway tells us that the duck is

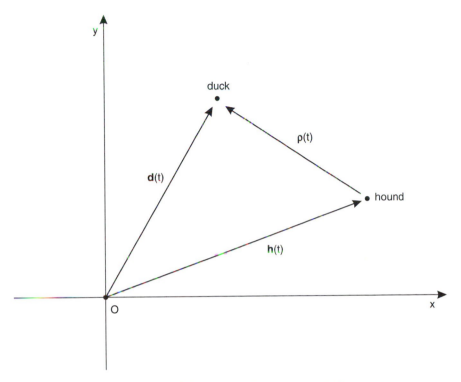

Figure 2.2.1 The geometry of the general hound-and-duck pursuit

always moving at a constant speed along a circle) is given by

$$\frac{d\mathbf{d}(t)}{dt} = \frac{dx_d}{dt} + i\frac{dy_d}{dt}$$

and so the *speed* of the duck is

$$\left|\frac{d\mathbf{d}(t)}{dt}\right| = \sqrt{\left(\frac{dx_d}{dt}\right)^2 + \left(\frac{dy_d}{dt}\right)^2}.$$

Since we are also told that the hound moves n times faster than does the duck, we have the speed of the hound as

$$n\sqrt{\left(\frac{dx_d}{dt}\right)^2 + \left(\frac{dy_d}{dt}\right)^2},$$

and so (2.2.4) becomes

$$(2.2.5) \qquad \frac{d\mathbf{h}(t)}{dt} = n\sqrt{\left(\frac{dx_d}{dt}\right)^2 + \left(\frac{dy_d}{dt}\right)^2} \cdot \frac{\boldsymbol{\rho}(t)}{|\boldsymbol{\rho}(t)|}$$

$$= n\sqrt{\left(\frac{dx_d}{dt}\right)^2 + \left(\frac{dy_d}{dt}\right)^2} \cdot \frac{\mathbf{d}(t) - \mathbf{h}(t)}{|\mathbf{d}(t) - \mathbf{h}(t)|}.$$

Writing (2.2.5) out in Cartesian coordinates gives

$$(2.2.6) \quad \frac{dx_h}{dt} + i\frac{dy_h}{dt} = n\sqrt{\left(\frac{dx_d}{dt}\right)^2 + \left(\frac{dy_d}{dt}\right)^2} \cdot \frac{(x_d - x_h) + i(y_d - y_h)}{\sqrt{(x_d - x_h)^2 + (y_d - y_h)^2}},$$

and, equating the real and imaginary parts of (2.2.6), we have the pair of differential equations

$$(2.2.7) \qquad \frac{dx_h}{dt} = n\sqrt{\left(\frac{dx_d}{dt}\right)^2 + \left(\frac{dy_d}{dt}\right)^2} \cdot \frac{x_d - x_h}{\sqrt{(x_d - x_h)^2 + (y_d - y_h)^2}},$$

and

$$(2.2.8) \qquad \frac{dy_h}{dt} = n\sqrt{\left(\frac{dx_d}{dt}\right)^2 + \left(\frac{dy_d}{dt}\right)^2} \cdot \frac{y_d - y_h}{\sqrt{(x_d - x_h)^2 + (y_d - y_h)^2}}.$$

Notice carefully that (2.2.7) and (2.2.8) are *very general*; unlike with (2.1.8) and (2.1.10), we have *at no time* made use of the assumption that the duck is swimming around a circle or that the hound starts his pursuit at the center of the duck's circle. Equations (2.2.7) and (2.2.8) are true for *any* path the duck might swim, as well as for the hound beginning its pursuit of the duck from *any* initial point (perhaps even from *outside* the duck's pond). If, however, we *do* make the assumption of the duck swimming around a circular path of unit radius (in whatever units of distance we care to use), at a speed such that it completes one revolution each 2π seconds (which of course means that the duck is swimming at a speed of one unit of distance per second), and if we further assume that at time $t = 0$ the duck is at the point $(1, 0)$, *then*

$$x_d(t) = \cos(t) \quad \text{and} \quad y_d(t) = \sin(t),$$

and so

$$n\sqrt{\left(\frac{dx_d}{dt}\right)^2 + \left(\frac{dy_d}{dt}\right)^2} = n\sqrt{\sin^2(t) + \cos^2(t)} = n.$$

Then (2.2.7) and (2.2.8) become the coupled differential equations

(2.2.9)
$$\frac{dx_h}{dt} = n\frac{\cos(t) - x_h}{\sqrt{(\cos(t) - x_h)^2 + (\sin(t) - y_h)^2}}$$

and

(2.2.10)
$$\frac{dy_h}{dt} = n\frac{\sin(t) - y_h}{\sqrt{(\cos(t) - x_h)^2 + (\sin(t) - y_h)^2}}.$$

These are the two equations for Hathaway's original problem that we can now solve on a computer, using, remember, *any* initial point for the hound's location at $t = 0$, about which we have made *no* assumption. If one were given some other swimming path for the duck, then the specific equations for $x_d(t)$ and $y_d(t)$ would simply be suitably modified and inserted into (2.2.7) and (2.2.8). This is of no consequence for a computer, of course, as specific analytic expressions are simply different grist for the same mill.[5]

This is not a computer programming book, and so I'm not going to get into the details of writing the computer code that solves (2.2.9) and (2.2.10). That code would be different for each programming language used anyway (in this book I use MATLAB®7.3's powerful differential equation solver named ode45[6]) — of course, the *results* produced by the various possible codes had better be the same! In figures 2.2.2 through 2.2.5 I have shown the pursuit curves that result from (2.2.9) and (2.2.10) when the hound, aka *dog* (the unlabeled circle), starts its pursuit at various points in the pond, including the center, Hathaway's original problem, with the counterclockwise swimming duck faster than the dog ($n = 0.3$). In figures 2.2.6 and 2.2.7 I started the hound

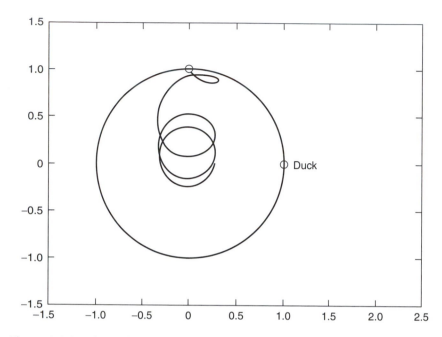

Figure 2.2.2 A hound-and-duck pursuit for $n = 0.3$

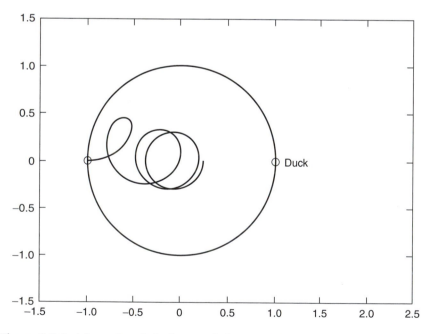

Figure 2.2.3 A hound-and-duck pursuit for $n = 0.3$

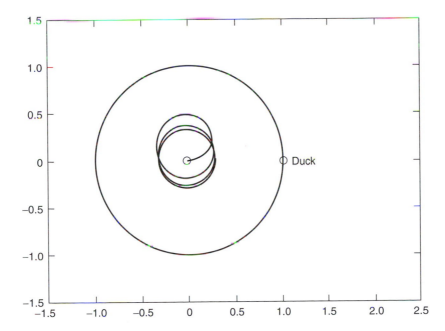

Figure 2.2.4 A hound-and-duck pursuit for $n = 0.3$

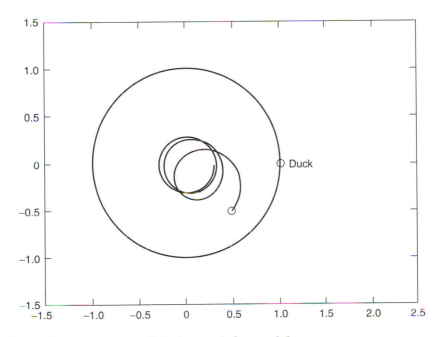

Figure 2.2.5 A hound-and-duck pursuit for $n = 0.3$

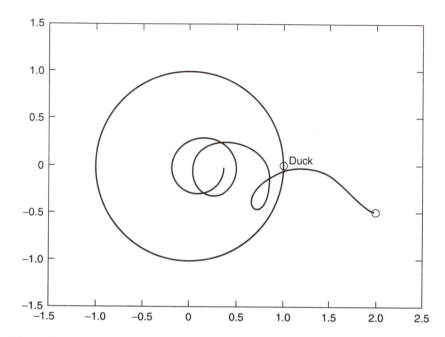

Figure 2.2.6 A hound-and-duck pursuit for $n = 0.3$

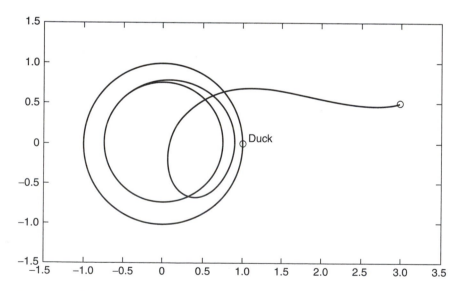

Figure 2.2.7 A hound-and-duck pursuit for $n = 0.75$

from outside the pond, once for $n = 0.3$ and again, from a different starting point, using $n = 0.75$. Despite all these variations, the plotted pursuit curves strongly hint at a remarkable general conclusion that is easy to understand in retrospect, but perhaps would go unnoticed if one were looking just at the differential equations (2.1.8) and (2.1.10): for $n < 1$ the path of the hound *always* becomes, as $t \to \infty$, a circle with radius na centered on the middle of the pond, where a is the radius of the pond. This general result appears, from our plots, to be independent of the starting point of the hound's pursuit. The asymptotic circle was given the name *limit cycle* by the French mathematician Henri Poincaré (1854–1912). Once we "see" this, I'll next show you that the circular limit cycle is no illusion, or even simply a coincidence for the particular values of n used in the plots of figures 2.2.2 through 2.2.7.

Writing $\varphi = \omega - \theta$ (take another look back at figure 2.1.2), we have $d\omega/dt = d\phi/dt + 1$, and so (2.1.8) becomes

$$(2.2.11) \qquad \rho \frac{d\phi}{dt} + \rho = a \cos(\varphi).$$

Also, (2.1.10) becomes

$$(2.2.12) \qquad \frac{d\rho}{dt} = a \sin(\phi) - an.$$

Differentiating (2.2.12) we have

$$\frac{d^2\rho}{dt^2} = a \cos(\phi) \frac{d\phi}{dt},$$

or

$$(2.2.13) \qquad \frac{d\phi}{dt} = \frac{d^2\rho/dt^2}{\cos(\varphi)}$$

which, when substituted in (2.2.11), gives

$$(2.2.14) \qquad \rho \frac{d^2\rho}{dt^2} + a\rho \cos(\phi) = a^2 \cos^2(\phi).$$

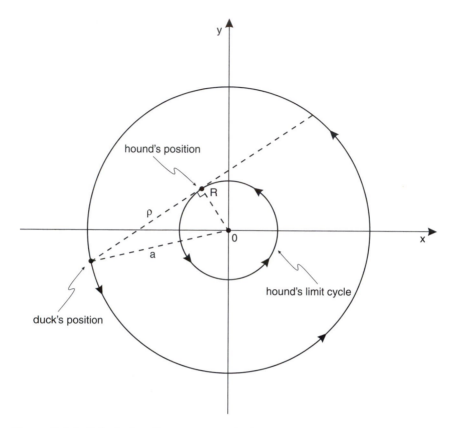

Figure 2.2.8 Calculating $\lim_{t\to\infty} \rho$ geometrically

Now, to see what is implied by (2.2.14), think about what being on a circular limit cycle for the hound *means* — it means that the line joining the hound on its circular path to the duck on *its* circular path is one-half of the chord that is *tangent* (don't forget, this is a pure pursuit problem) to the inner circle (the point of tangency is the hound's position) and connects to the outer circle at the duck's position, as shown in figure 2.2.8. As the hound and duck move, the chord joining them rotates, *but its length remains unchanged*. This means that $\lim_{t\to\infty} \rho$ is a constant, and so, once the path of the hound has settled down to its circular limit cycle, it must be true that $d\rho/dt = d^2\rho/dt^2 = 0$. We can calculate the value of $\lim_{t\to\infty} \rho$ as follows. From (2.2.11) we see that, as $t \to \infty$, $\rho = a\cos(\phi)$, and from (2.2.12) we see that, as $t \to \infty$,

$\sin(\phi) = n$. Substituting all these conclusions into (2.2.14), we see that as $t \to \infty$

$$a\rho\left(\frac{\rho}{a}\right) = a^2[1 - \sin^2(\phi)] = a^2(1 - n^2),$$

or

$$\rho^2 = a^2(1 - n^2),$$

or

(2.2.15) $$\lim_{t \to \infty} \rho = a\sqrt{1 - n^2}.$$

So, calling the radius of the limit cycle R, the Pythagorean theorem says

$$R^2 + \rho^2 = a^2,$$

or

$$R = \sqrt{a^2 - \rho^2} = \sqrt{a^2 - a^2(1 - n^2)},$$

or, as claimed earlier,

(2.2.16) $$R = na.$$

As a final comment on Hathaway's problem, after playing for a while with computer solutions as a function of the hound's starting position, it becomes clear that, if the starting point is *anywhere in the pond*, then, *if* capture occurs it will be from *behind* the duck. For example, even for a "fast" hound ($n = 1.5$) starting "above and in front" of the duck but still in the pond, as shown in figure 2.2.9, the hound captures the duck only after swimming in around and behind the duck with the asterisk marking the end of the pursuit. If the hound starts from *outside* the pond, however, as Bernhart (1953) says, it is possible that "the duck will swim directly into the dog's mouth!" Figure 2.2.10 shows such a frontal capture.

Now, to end this section here's a challenge problem for you to consider. Three years after Hathaway's 1921 paper appeared, a reply of a sort was published — Morley (1924) — written by Frank Morley (1860–1937), former president of the American Mathematical Society (1919–1920), a professor of mathematics at The Johns Hopkins

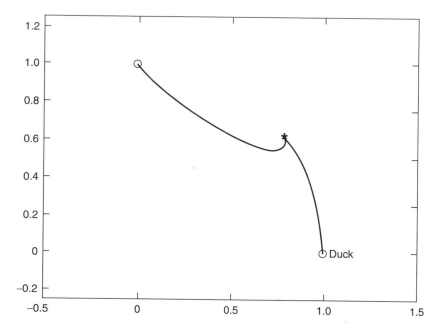

Figure 2.2.9 Capture from behind

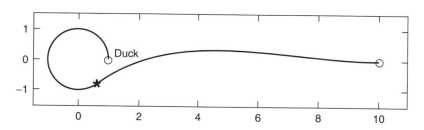

Figure 2.2.10 Frontal capture

University, and the father of F. V. Morley, who was the author of Morley (1921) that I cited back in chapter 1. After commenting on the difficulty of the mathematics when the *pursued* moves on a circular path, the senior Morley wrote:

> It occurred to me that the case of a dog D running around a man M might be more tractable. What is meant is that the path of D is always at right angles to the line DM. For a given locus of M [the curve for D] may be called a *curve of ambience*.

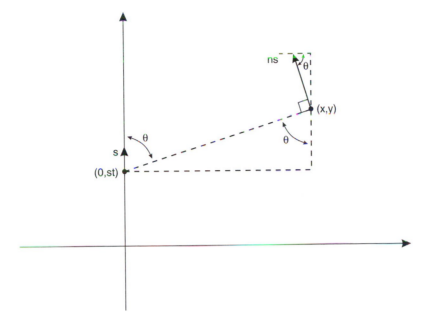

Figure 2.2.11 The geometry of Morley's running dog problem

Morley's name makes sense since, in general, the dog (if faster than the man) will "run around" the man. Textbook writers, curiously, seem never to have picked up on Morley's problem, the calculation of the dog's path. That may be due to the manner in which Morley presented his analysis, which is (in my opinion) extraordinarily dense and difficult to understand. So, your problem here is to attack Morley's problem in a way completely different from his approach, but quite similar to the way Bouguer's original pirate-merchant pursuit was analyzed in the first chapter. Indeed, before tackling this challenge problem you may want to take another look at section 1.1.

For this challenge problem we'll limit ourselves to a fast dog ($n > 1$), and initially take the man and the dog, respectively, at (0, 0) and (1, 0). The dog starts running in the counterclockwise sense at time $t = 0$ and, at some arbitrary time $t > 0$, the man (moving at speed s up the y-axis) and the dog (running at speed ns) will be located as shown in figure 2.2.11. In the notation of that figure, you should see that the differential equations describing the dog's motion are

$$\frac{dx}{dt} = -ns \cos(\theta) \quad \text{and} \quad \frac{dy}{dt} = ns \sin(\theta),$$

and so

(2.2.17)
$$\frac{dy}{dx} = -\tan(\theta) = \frac{x}{st - y},$$

when the dog is at (x, y) at time t. So, here's your problem: find the equation of the dog's path $y = y(x)$ for $0 \le x \le 1$ and, in particular, find an expression for the value of $y(0)$, the ordinate of the dog at the point at which it first runs directly across (from right to left) the path of the man. Both of your expressions will, of course, be a function of n. If you want to check your answers, you'll find solutions worked out in appendix D.

2.3 Velocity and Acceleration Calculations for a Moving Body

In this section I want to show you a clever way to derive quickly the general formulas for the velocity and acceleration of a body moving in a plane. We'll use these formulas in later sections of the book when discussing maneuvering targets of pursuit. The derivation of these formulas, using complex quantities, requires only that one know how to differentiate an exponential, and that multiplying a vector in the complex plane by $i = \sqrt{-1}$ *rotates* that vector counterclockwise through 90°. If you know what a differential equation is — the basic assumption of this book — then I think I can safely assume these additional two pieces of mathematics aren't new to you, either.

In figure 2.3.1 I have drawn the position vector $\mathbf{p}(t)$ of a body located, at time t, at the point (x, y). The polar coordinates of the body are (r, θ), where, of course,

$$x = r \cos(\theta), \quad y = r \sin(\theta).$$

From Euler's identity, we can write these equations as

$$\mathbf{p}(t) = x + iy = r \cos(\theta) + ir \sin(\theta) = re^{i\theta}.$$

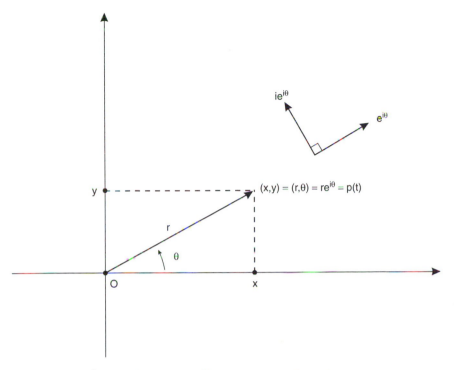

Figure 2.3.1 The position vector of a moving body in a plane

Be sure to understand what that benign-looking exponential *is*: $e^{i\theta}$ is a *radial vector* of magnitude (length) *one* pointing directly to the body at angle θ with respect to the horizontal axis, while r is a *scalar* amplitude. Now, the velocity vector $\mathbf{v}(t)$ of the body is the time derivative of the position vector, and the body's acceleration vector $\mathbf{a}(t)$ is the time derivative of the velocity vector. Both θ and r are generally functions of time. So,

$$\mathbf{v}(t) = \frac{d}{dt}\mathbf{p}(t) = \frac{dr}{dt}e^{i\theta} + re^{i\theta}i\frac{d\theta}{dt}$$

and

$$\mathbf{a}(t) = \frac{d}{dt}\mathbf{v}(t)$$

$$= \left(\frac{d^2r}{dt^2}e^{i\theta} + \frac{dr}{dt}e^{i\theta}i\frac{d\theta}{dt}\right)$$

$$+ i\left(re^{i\theta}\frac{d^2\theta}{dt^2} + \frac{dr}{dt}e^{i\theta}\frac{d\theta}{dt} + rie^{i\theta}\frac{d\theta}{dt}\cdot\frac{d\theta}{dt}\right).$$

That is,

(2.3.1)
$$\mathbf{v}(t) = \frac{dr}{dt}(e^{i\theta}) + r\frac{d\theta}{dt}(ie^{i\theta})$$

and

(2.3.2) $$\mathbf{a}(t) = \left[\frac{d^2r}{dt^2} - r\left(\frac{d\theta}{dt}\right)^2\right](e^{i\theta}) + \left[2\frac{dr}{dt}\cdot\frac{d\theta}{dt} + r\frac{d^2\theta}{dt^2}\right](ie^{i\theta}).$$

Keeping in mind that $e^{i\theta}$ is a unit (length) *radial* vector, we can understand why $ie^{i\theta}$ is called a unit (length) *transverse* vector — it is just $e^{i\theta}$ rotated counterclockwise by 90°, as shown in figure 2.3.1. Therefore, from (2.3.1) and (2.3.2) we have the following results:

- the moving body has the *radial* velocity component $\frac{dr}{dt}$, and the *transverse* (or *lateral*) velocity component $r d\theta/dt$;
- the moving body has the *radial* acceleration component $d^2r/dt^2 - r(d\theta/dt)^2$, and the *transverse* (or *lateral*) acceleration component $2(dr/dt)\cdot(d\theta/dt) + r d^2\theta/dt^2$.

As an example of the usefulness of this last formula, we can use it to derive a result of immense historical importance in astronomical physics. The planets of the solar system orbit the sun under the influence of what is called a *central force field* (gravity!), which mean that the gravitational force of the sun on a planet is always strictly directed toward the center of the sun. *That* means, because of Newton's famous "$F = ma$," the acceleration of a planet is strictly radial, with *zero* transverse acceleration. Thus, from our last result,

$$2\frac{dr}{dt}\cdot\frac{d\theta}{dt} + r\frac{d^2\theta}{dt^2} = 0,$$

where r is the distance between the centers of the sun and the planet and θ is the angle made by the radius vector from the sun and the planet with respect to some (arbitrary) reference line in the plane of the planet's orbit. Now let $u = d\theta/dt$, and so

$$2\frac{dr}{dt}u + r\frac{du}{dt} = 0,$$

or

$$2u\,dr + r\,du = 0,$$

or

$$\frac{du}{u} = -2\frac{dr}{r}.$$

Integrating indefinitely, with C_1 as the integration constant, we have

$$\ln(u) = -2\ln(r) + C_1$$

or

$$\ln(u) + \ln(r^2) = \ln(ur^2) = C_1$$

or

$$ur^2 = e^{C_1} = C,$$

where C is some constant, and so

$$r^2\frac{d\theta}{dt} = \text{constant.}$$

To see what this means physically, look at figure 2.3.2, which shows the planet's location at times t and $t + \Delta t$, with the sun at the origin of our coordinate system. We assume Δt is so small that the angular change $\Delta\theta$ in the radius vector's angle is also very small, as well as that the length of the radius vector remains essentially unchanged. Then the shaded area is virtually that of an isosceles triangle with *differential* area ΔA given by

$$\Delta A = \frac{1}{2}\text{base} \times \text{height} \approx \frac{1}{2}(r\,\Delta\theta)r = \frac{1}{2}r^2\Delta\theta.$$

Dividing through by Δt gives

$$\frac{\Delta A}{\Delta t} \approx \frac{1}{2}r^2\frac{\Delta\theta}{\Delta t},$$

an expression that becomes exact as $\Delta t \to 0$: that is, replacing Δt with dt, ΔA with dA, and $\Delta\theta$ with $d\theta$, we have

$$\frac{dA}{dt} = \frac{1}{2}r^2\frac{d\theta}{dt}.$$

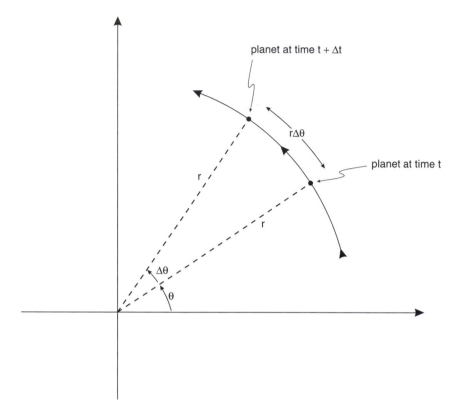

Figure 2.3.2 Interpreting $r^2\frac{d\theta}{dt}$

Thus

$$r^2\frac{d\theta}{dt} = 2\frac{dA}{dt}$$

and so our earlier result says

$$\frac{dA}{dt} = \text{constant,}$$

or, in words, the radius vector connecting the sun and the planet sweeps out area at a constant rate. Or, as the German astronomer Johann Kepler (1571–1630) deduced from direct, *observational* data, *the radius vector sweeps over equal areas in equal intervals of time*, a statement famously known today as Kepler's second law of planetary motion.

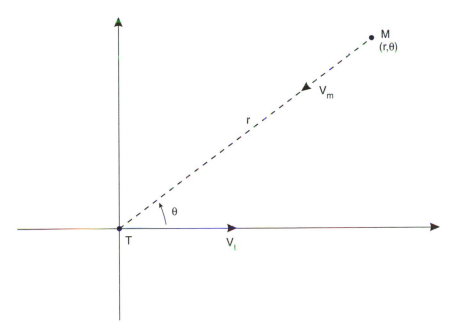

Figure 2.3.3 Homing missile attack geometry

As a *pursuit* example of the use of our general velocity and acceleration formulas, consider the following problem, with reference to figure 2.3.3. A missile M is in pure pursuit of the nonmanuevering target T — M and T are at the same constant altitude above the earth, and so the pursuit takes place in a plane — with constant speeds V_m and V_t, respectively. (When a *pure* pursuit missile is employed, missile engineers often call their gadget a *homing* missile.) It is assumed that the target flies in a straight line. The instant of the missile launch defines time $t = 0$, and the launch occurs at a range of $r(t = 0) = R_0$. If we use the target's initial position as our coordinate origin, and observe that we can always orient the coordinate axes so that the target flies along the positive horizontal axis, then the missile's location in polar coordinates is (r, θ), where $\theta(t = 0) = \theta_0$; θ_0 is called the *missile launch angle*. For example, $\theta_0 = 0$ would represent a "head-on" missile launch, $\theta_0 = \pi/2$ would represent a "broadside" launch, and $\theta_0 = \pi$ would represent a "tail chase" launch. While "head-on" and "tail chase" launches are perfectly feasible scenarios, the pure pursuit of a nonmanuevering target in these two special cases is not of much mathematical interest. So, we'll limit our analyses to $0 < \theta_0 < \pi$ when we calculate the following two things: (1) the general path of the missile and (2) the required acceleration of the missile *just before* impact with the target (the so-called *terminal* acceleration).

Before we do those calculations, however, let me address one matter that may now be bothering you. It is one thing for a dog, or a hawk, or a human to execute a pure pursuit strategy, or course, as the means for doing so are obvious — just use your eyes! But how does a *missile* do it? A common approach is based on what is called a *passive infrared* (IR) *seeker*, where *passive* means no missile-generated signals are used (as would be done when using an *active* seeker, such as radar). The passive IR sensor is made from a special material that generates an electrical signal when radiation of a certain wavelength falls on it; the amplitude of the electrical signal is proportional to the intensity of the incident radiation. For a missile seeker the wavelength of interest is in the infrared region of the electromagnetic spectrum, where the heat radiation emitted by hot jet or rocket engines is located. The IR sensor is fastened to a three-degree-of-freedom gymbal mechanism powered by electrical motors that, depending on the electrical input signals to them, drive the gymbal and the IR sensor — the two together are located in the nose of the missile (looking forward) in what is called the *seeker cage* — up, down, and around in space. Think of an "electronic eye flying through the sky!"

The gymbal drive motor input signals are generated by the missile guidance electronics package in such a way as to maintain the *maximum* IR sensor output signal (the gymbal is said to be *locked onto* the target) — this is a classic example of an electronic *feedback system*. Given the instantaneous orientation of the seeker cage, additional electronic circuitry senses the deviation between the instantaneous body-axis of the cage and the instantaneous tail-to-nose body-axis of the missile. That deviation is transformed into corrective electrical signals that control the aerodynamic flight control surfaces of the missile so as to generate forces (accelerations) that tend to align the two body axes, that is, to drive the deviation to zero — *another* feedback system.

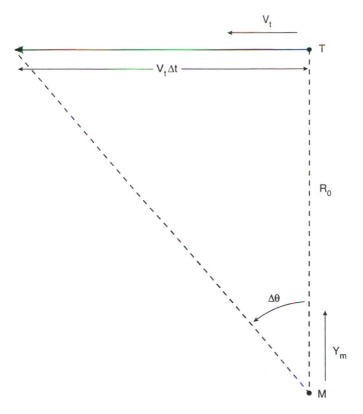

Figure 2.3.4 Calculating the initial angular rate-of-change of the line-of-sight from a missile to a broadside target

One important parameter for such a missile guidance system is the so-called *seeker cage slew rate*, which is how fast, in degrees/second, the seeker cage can swing in space as it attempts to track a target. For any real system this number clearly must always be finite. As an elementary example of a typical calculation of the slew rate, take a look at figure 2.3.4, which shows a missile being launched broadside to a target flying in a straight line at speed V_t at initial range R_0. In time interval Δt the target moves a distance of $V_t \Delta t$, and so the line-of-sight angle from missile to target changes by $\Delta\theta$; thus

$$\frac{V_t \Delta t}{R_0} = \tan(\Delta\theta).$$

As $\Delta t \to 0$ we have $\Delta \theta \to 0$ and so, writing $d\theta$ for $\Delta \theta$ and using $\tan(\Delta \theta) \approx \Delta \theta$ for $\Delta \theta$ "small,"

$$\frac{V_t}{R_0} dt = d\theta,$$

or, at the instant of launch,

$$\frac{d\theta}{dt} = \frac{V_t}{R_0}.$$

If $R_0 = 25,000$ feet and $V_t = 1,000$ feet/second, then the initial value of $d\theta/dt$ at launch is

$$\frac{d\theta}{dt} = \frac{1,000}{25,000} = \frac{1}{25} \text{ radians/second} = 2.3 \text{ degrees/second.}$$

This is well within the seeker cage slew rates of modern air-to-air intercept missiles; for example, the decades-old Hughes/Raytheon Defense Systems AIM-9J Sidewinder has a seeker cage slew rate of 16 degrees/second.

From figure 2.3.3 we can see, geometrically, that if we resolve the missile velocity vector (which always points directly at the target's instantaneous location) into radial and transverse components, along the line-of-sight from M to T and along a direction perpendicular to the line-of-sight, respectively, then from our earlier results for the velocity components we have

$$(2.3.3) \qquad \frac{dr}{dt} = -V_m - V_t \cos(\theta) = -[V_m + V_t \cos(\theta)]$$

and

$$(2.3.4) \qquad r\frac{d\theta}{dt} = V_t \sin(\theta).$$

The signs of dr/dt and $d\theta/dt$ are obviously correct from the physics of the pure pursuit, that is, $dr/dt < 0$ (the missile continually gets *closer* to the target) and $d\theta/dt \geq 0$ (the line-of-sight angle from the target to the missile continually *increases*) over the duration of the entire pursuit, from launch to impact.

Dividing (2.3.3) by (2.3.4), we get

$$\frac{dr/dt}{rd\theta/dt} = \frac{-[V_m + V_t \cos(\theta)]}{V_t \sin(\theta)},$$

or, if we define the constant $k = V_m/V_t$,

(2.3.5)
$$-\frac{dr}{r} = \frac{k + \cos(\theta)}{\sin(\theta)} d\theta.$$

Integrating indefinitely (using a good table of integrals),

$$-\ln(r)\,|_{R_0}^r = k \int \frac{d\theta}{\sin(\theta)} + \int \cot(\theta) d\theta = k\ln\left\{\tan\left(\frac{\theta}{2}\right)\right\} + \ln\{\sin(\theta)\} + \ln(C),$$

where $\ln(C)$ is the constant of indefinite integration, and so

$$\ln\left(\frac{R_0}{r}\right) = \ln\left\{C \sin(\theta) \tan^k\left(\frac{\theta}{2}\right)\right\},$$

or

(2.3.6)
$$\frac{R_0}{r} = C \sin(\theta) \tan^k\left(\frac{\theta}{2}\right).$$

Since $r = R_0$ when $\theta = \theta_0$, then

$$C = \frac{1}{\sin(\theta_0) \tan^k(\theta_0/2)},$$

or, for our first result, the polar equation of the flight path of the missile (remember, the coordinate system origin is the *target's* location at time $t = 0$),

(2.3.7)
$$\boxed{r = R_0 \frac{\sin(\theta_0)}{\sin(\theta)} \cdot \left\{\frac{\tan(\theta_0/2)}{\tan(\theta/2)}\right\}^k, \quad k = \frac{V_m}{V_t}.}$$

Since $\sin(\theta) = 2\sin(\theta/2)\cos(\theta/2)$, then

$$r = R_0 \frac{2\sin(\theta_0/2)\cos(\theta_0/2)}{2\sin(\theta/2)\cos(\theta/2)} \cdot \left\{ \frac{\sin(\theta_0/2)/\cos(\theta_0/2)}{\sin(\theta/2)/\cos(\theta/2)} \right\}^k,$$

or

$$r = R_0 \frac{\sin^{k+1}(\theta_0/2)/\cos^{k-1}(\theta_0/2)}{\sin^{k+1}(\theta/2)/\cos^{k-1}(\theta/2)} = R_0 \left\{ \frac{\cos(\theta/2)}{\cos(\theta_0/2)} \right\}^{k-1} \cdot \left\{ \frac{\sin(\theta_0/2)}{\sin(\theta/2)} \right\}^{k+1}.$$

(2.3.8)

As a quick partial check on this result, notice that *at launch* $\theta = \theta_0$ and (2.3.8) reduces to $r = R_0$, just as it should. More interesting, however, is that (2.3.8) tells us what θ becomes in the *terminal phase* of the missile's approach to the target, that is, what θ is just before impact. We know $r \to 0$ as impact is approached — by the very definition of what *impact* means — and this can only occur (in the $k > 1$ case, the case of practical interest) by having $\cos(\theta/2) \to 0$, that is, $\theta \to \pi$. In other words, the missile *always* ends up impacting from *behind* the target, *independent* of the initial launch angle θ_0 (for all $\theta_0 \neq 0$) as well as independent of the missile's speed no matter how large that may be. I don't think this is intuitively obvious, at all.

Next, as a prelude to calculating the answer to our second problem (what is the *terminal* acceleration of the missile?), return to (2.3.4) to write

$$\frac{d\theta}{dt} = \frac{V_t \sin(\theta)}{r}$$

or, using (2.3.8),

$$(2.3.9) \qquad \frac{d\theta}{dt} = \frac{V_t}{R_0} \sin(\theta) \left\{ \frac{\cos(\theta_0/2)}{\cos(\theta/2)} \right\}^{k-1} \cdot \left\{ \frac{\sin(\theta/2)}{\sin(\theta_0/2)} \right\}^{k+1}.$$

Again using the identity $\sin(\theta) = 2\sin(\theta/2)\cos(\theta/2)$, we see that (2.3.9) becomes

$$\frac{d\theta}{dt} = \frac{V_t}{R_0} \cdot 2\sin\left(\frac{\theta}{2}\right)\cos\left(\frac{\theta}{2}\right) \cdot \left\{\frac{\cos(\theta_0/2)}{\cos(\theta/2)}\right\}^{k-1} \cdot \left\{\frac{\sin(\theta/2)}{\sin(\theta_0/2)}\right\}^{k+1}$$

$$= \frac{2V_t}{R_0} \cdot \frac{\cos^{k-1}(\theta_0/2)}{\sin^{k+1}(\theta_0/2)} \cdot \frac{\sin^4(\theta/2)\sin^{k-2}(\theta/2)}{\cos^{k-2}(\theta/2)},$$

or

$$(2.3.10) \qquad \frac{d\theta}{dt} = 2\frac{V_t}{R_0} \cdot \frac{\cos^{k-1}(\theta_0/2)}{\sin^{k+1}(\theta_0/2)} \cdot \sin^4\left(\frac{\theta}{2}\right) \cdot \tan^{k-2}\left(\frac{\theta}{2}\right).$$

Since $\theta \to \pi$ as the missile approaches impact for all $k \geq 1$ then, since $\sin(\pi/2) = 1$ and $\tan(\pi/2) = \infty$, we have

$$\lim_{\theta \to \pi,\, \theta_0 \neq 0} \frac{d\theta}{dt} = \begin{cases} \infty, & \text{for } k > 2, \\ 0, & \text{for } 1 < k < 2. \end{cases}$$

When $k = 2$,

$$\frac{d\theta}{dt} = 2\frac{V_t}{R_0} \cdot \frac{\cos(\theta_0/2)}{\sin^3(\theta_0/2)} \cdot \sin^4\left(\frac{\theta}{2}\right),$$

and so

$$\lim_{\theta \to \pi,\, \theta_0 \neq 0} \frac{d\theta}{dt} = 2\frac{V_t}{R_0} \cdot \frac{\cos(\theta_0/2)}{\sin^3(\theta_0/2)}, \quad k = 2.$$

Continuing with our prelude, notice that since the missile and the target are each *constant speed* objects (zero acceleration along the tail-to-nose body axis) then, just before impact, the *tail chase nature* of the interception means (besides, of course, $r \to 0$) that

$$(2.3.11) \qquad \lim_{\theta \to \pi} \frac{dr}{dt} = V_t - V_m = V_t\left(1 - \frac{V_m}{V_t}\right) = V_t(1 - k),$$

as well as

$$\lim_{\theta \to \pi} \frac{d^2r}{dt^2} = 0.$$

For $1 < k < 2$, these results so far show that the missile's *terminal* radial acceleration is

$$\lim_{\theta \to \pi} \left[\frac{d^2r}{dt^2} - r \left(\frac{d\theta}{dt} \right)^2 \right] = 0.$$

For the transverse acceleration, we need to know $d^2\theta/dt^2$. Looking back at (2.3.10), we see that

(2.3.12) $$\frac{d\theta}{dt} = C \sin^4 \left(\frac{\theta}{2} \right) \cdot \tan^{k-2} \left(\frac{\theta}{2} \right)$$

where C is a constant. For $1 < k < 2$ this means that

$$\frac{d\theta}{dt} = C \sin^4 \left(\frac{\theta}{2} \right) \cdot \frac{\cos^{2-k}(\theta/2)}{\sin^{2-k}(\theta/2)} = C \sin^{k+2} \left(\frac{\theta}{2} \right) \cos^{2-k} \left(\frac{\theta}{2} \right),$$

and so

$$\lim_{\theta \to \pi} \frac{d\theta}{dt} = 0, \, 1 < k < 2,$$

and it should also be obvious, by inspection, that

$$\lim_{\theta \to \pi} \frac{d^2\theta}{dt^2} = 0, \, 1 < k < 2.$$

From all this we can now write the missile's *terminal* $(r \to 0)$ transverse acceleration as

$$\lim_{\theta \to \pi} \left[2\frac{dr}{dt} \cdot \frac{d\theta}{dt} + r\frac{d^2\theta}{dt^2} \right] = 0, \, 1 < k < 2.$$

So, for $1 < k < 2$ the *total* terminal missile acceleration is zero.

Next, suppose that $k > 2$. Let's start by calculating the terminal transverse acceleration. From (2.3.11) and (2.3.12) we see that the first term in that acceleration is

$$\lim_{\theta \to \pi} 2\frac{dr}{dt} \cdot \frac{d\theta}{dt} = \lim_{\theta \to \pi} 2V_t(1 - k)C \sin^4\left(\frac{\theta}{2}\right) \tan^{k-2}\left(\frac{\theta}{2}\right) = -\infty$$

if $k > 2$. The second term of the transverse acceleration is also clearly *negative*, as well, as $r > 0$ and $d^2\theta/dt^2 < 0$ (this last condition immediately follows, without the need for a detailed calculation, from the fact that θ increases with time to its asymptotic upper limit of π, i.e., $\theta(t)$ is concave downward and so while $d\theta/dt > 0$ always, the value of $d\theta/dt$ is always *decreasing* and we have

$$\frac{d}{dt}\left(\frac{d\theta}{dt}\right) = \frac{d^2\theta}{dt^2} < 0.$$

Thus, $r\, d^2\theta/dt^2 < 0$. And so the total transverse acceleration is (negative) infinity. With this result we don't need to bother with doing a radial acceleration calculation — we can immediately conclude that the total missile acceleration is, for $k > 2$, *infinite*.

Since the forces that must be applied by the missile's air-flow control surfaces, for example, wing and/or tail fins, increase with the desired acceleration (the missile must make a "tighter" turn in and around onto the target's tail in the terminal phase as the missile's speed increases), there will always be some maximum speed limit beyond which the missile's control systems will be unable to generate the required forces to properly move the wing and/or tail fin surfaces. I do think it is, however, a completely surprising result that the acceleration of a missile performing pure pursuit is, just before impact, a *discontinuous* function of k. It is interesting to note that the observation of a discontinuity in the terminal missile acceleration occurs at $k = 2$ didn't appear in the technical literature[7] until the relatively recent date of 1948. The immediate practical implication that follows from all this is that "fast" missiles (missiles that travel faster than twice the speed of their targets) can**not** employ a pure pursuit strategy all the way from start (launch) to finish (impact). And, in fact, a different pursuit strategy called *proportional navigation* is the actual strategy that is generally used in practice by real, operational missiles attacking moving targets. That strategy will be discussed in the last section of this chapter.

Now, here's a challenge problem for you to end this section. Consider the special case of $k = 2$ for a pure pursuit missile. You are to do the following three calculations:

(1) Derive an expression for the missile's terminal acceleration and thereby show that it is greater than zero but *finite*, and numerically evaluate it for the data ($R_0 = 25,000$ feet, $V_t = 1,000$ feet/second, and $\theta_0 = 90°$, a broadside launch). Express your answer in "gees," where one gee is the acceleration of gravity at the earth's surface (32 feet/second^2).

(2) Calculate the total flight time for a broadside launch, from launch to impact, in terms of R_0 and V_m ($= 2V_t$).

(3) Redo the calculation of (2) for an initial launch angle of $\theta_0 = 60°$ (a combination head-on and broadside launch). Before you start, do you think the flight time will now be greater than or less than (for the same R_0 and V_m, of course) than the flight time derived in (2)? Does your calculation agree with your intuition?

You can compare your answers with my solutions in appendix E.

2.4 Houghton's Problem: A Circular Pursuit That *Is* Solvable in Closed Form

In an essay I've cited before that discusses Hathaway's circular pursuit problem — Archibald and Manning (1921) — there is a provocative footnote that mentions an undated five-page pamphlet devoted to the $n = 1$ case (dog and duck moving with equal speeds). Titled *A Common Sense Solution of a Curve of Pursuit Problem That Has Been Considered Unsolvable by Many Eminent Mathematicians*, the author was given as one L. T. Houghton of Worcester, Mass. (of whom I've been able to learn nothing). Unlike Hathaway, who declared that the $n = 1$ case would *not* result in a capture, Houghton asserted capture *would* occur. Houghton had apparently been long engaged in spirited debates with others over his claim because, at one point in his pamphlet, he wrote

This curve of pursuit problem has estranged old friends and vexed eminent mathematicians. Many wagers have been made which professors of mathematics have been called on to settle, and their decisions have been in the negative [that is, in agreement with Hathaway that, for equal speeds, the dog will *not* catch the duck], without one single line of proof to sustain their findings. A bare assertion is not satisfactory proof.

We can see, however, from (2.1.10), why Hathaway was correct. That equation says that the rate of change of ρ (the distance between the dog and the duck) with θ is, *at most*, zero. That most positive value of $d\rho/d\theta$ occurs when the dog is on its limit cycle, a circle with the same radius as the duck's circle of motion; that is, the dog is directly behind the duck and *never reduces the separation because they are moving at the same speed*. The origin of Houghton's error lies in his interpretation of the nature of the pursuit curve; he denied that the tangent to the dog's pursuit curve is always pointed at the duck, but claimed that the words "the dog always moves toward the duck" means that the center of the pond, the dog, and the duck, are always collinear. On that basis (which I personally find a bit tortured) Houghton asserted that the dog would capture the duck after the duck had swum through a distance equal to the pond's diameter, that is, through an arc subtending two radians ($\approx 114.6°$).

As the footnote in the *American Mathematical Monthly* pointed out, Houghton had simply attempted to solve a problem different from Hathaway's, and it is no surprise that he got a different answer. I say *attempted* because, to rub salt in the wound, it was also pointed out that even with Houghton's curious interpretation of the dog-duck pursuit his answer is wrong! The correct answer for the problem of a collinear pond center, dog, and duck, is that the dog catches the duck after the duck has swum through an arc of 90°. Houghton's interpretation is now often found in textbooks — but never, so far as I know, with attribution — with the dog and duck suitably altered to fit modern times.

One textbook author, for example, put it this way:

> A target moves on the circumference of a circle with constant speed V. A missile starts at the center of this circle and pursues the target. The speed of the missile is also V. The pursuit is such that the center of the circle, the missile, and the target are collinear. Show that the target moves through one-fourth of the circumference up to the moment of capture [a better word, I think, in this context, would be *impact*].

In this form the problem is a bit odd — what's the point of a missile keeping itself collinear with a target and the center of the target's circular path? This assumes the missile somehow knows the target will be flying a circular path and, in fact, *what* circular path, because we are told the missile knows where the center of that circle is. This requires sophisticated *global* knowledge, as opposed to the much more limited *local* knowledge which is all that is required by a pure pursuit strategy; in that strategy all the missile has to know is where the target is *now*. The reason for such an oddly stated problem, I think, is that it is simply a nice mathematical question that the author tried to dress up in modern language. A far better physical context is provided by the situation of a hunting dog pursuing a rabbit. A rabbit hunter typically has a dog with him to hunt for rabbit *holes* — when (if) the rabbit appears on the scene after a hard day of doing whatever a rabbit does, and spots the dog, it will typically run in a circle around its hole, hoping to lure the dog into a pure pursuit chase. If that strategy works, then the rabbit can suddenly make a dash straight for the safety of its hole. But that dash will fail and the rabbit will be defeated if the dog keeps itself, the rabbit, and the hole (the location of which, of course, the dog knows) *collinear*. That is a physically meaningful context for Houghton's problem. Well, perhaps this *is* all a bit pedantic and so, despite all the above, in the interest of scholarly fellowship I'll retain the missile/target language for the discussion that follows.

What I call *Houghton's problem* (the man does, after all, deserve *some* credit) is, in fact, easily solved. Indeed, let's be just a bit more general and solve it for the case where the missile moves at speed kV, $k \geq 1$, with V the speed of the target. With reference to the notation of figure 2.4.1, let's take the target's circular path to have a radius R.

We'll take the target to be at $(R, 0)$ at time $t = 0$, and in general the position vector of the target to be $\mathbf{p}(t) = Re^{i\theta(t)}$, where $\theta(t)$ is the angle the target's position vector makes with the horizontal axis (and $i = \sqrt{-1}$, of course). Since the target moves at speed V, it travels once around its circle, that is, through 2π radians, in $2\pi R/V$ seconds. Thus,

$$\frac{d\theta}{dt} = \frac{2\pi}{2\pi R/V} = \frac{V}{R},$$

and so $\theta(t) = Vt/R$. Thus, the target's position vector is

(2.4.1) $$\mathbf{T}(t) = Re^{iVt/R}.$$

The position vector of the missile will then be given by

(2.4.2) $$\mathbf{M}(t) = r(t)e^{i\theta(t)},$$

where $r(t)$ is some increasing function of t, with $r(0) = 0$ and $r(\hat{t}) = R$ if $\hat{t}$ denotes the time of impact. The angle of $\mathbf{M}(t)$ is, of course, the same $\theta(t)$ as in $\mathbf{T}(t)$, since the missile, target, and the center of the target's circular path are *given* as collinear.

The missile has a speed component $V_r(t)$ in the radial direction and a perpendicular speed component $V_\theta(t)$ in the transverse direction, where the Phythagorean theorem tells us that

(2.4.3) $$kV = \sqrt{V_r^2 + V_\theta^2}.$$

For the missile and the target always to be collinear with the center of the target's circular path, it is geometrically clear (look at figure 2.4.1 again) that

(2.4.4) $$V_\theta(t) = \frac{r(t)}{R}V.$$

Combining (2.4.3) and (2.4.4) gives us

$$k^2V^2 = V_r^2 + \frac{r^2}{R^2}V^2,$$

or

$$(2.4.5) \qquad V_r = V\sqrt{k^2 - \frac{r^2}{R^2}}, \ k \geq 1.$$

Now, since V_r is the *radial* speed, then $V_r = dr/dt$, and so

$$(2.4.6) \qquad \frac{dr}{dt} = V\sqrt{k^2 - \frac{r^2}{R^2}}.$$

The answer to Houghton's problem is simply the result of calculating how long it takes for the missile to travel from $r = 0$ to $r = R$. This can be found by integrating (2.4.6), because

$$dt = \frac{dr}{V\sqrt{k^2 - r^2/R^2}},$$

and then

$$\int_0^{\hat{t}} dt = \hat{t} = \frac{1}{V}\int_0^R \frac{dr}{\sqrt{k^2 - r^2/R^2}} = \frac{1}{kV}\int_0^R \frac{dr}{\sqrt{1 - r^2/k^2 R^2}}.$$

Changing variable to $x = r/kR$ (so $dr = kR\,dx$), we have

$$\hat{t} = \frac{1}{kV}\int_0^{1/k} \frac{kR\,dx}{\sqrt{1 - x^2}} = \frac{R}{V}\sin^{-1}(x)\,|_0^{1/k} = \frac{R}{V}\sin^{-1}\left(\frac{1}{k}\right).$$

In particular, if $k = 1$ then

$$\hat{t} = \frac{R}{V}\sin^{-1}(1) = \frac{\pi R}{2V} = \frac{1}{V}\cdot\frac{2\pi R}{4},$$

which is the time required for the target to travel along one-quarter of the circumference of its circular path, that is, through an arc of 90°, significantly less than Houghton's answer. Unlike pure pursuit, where there is no capture unless the missile is *strictly faster* than the target, there *is* capture in Houghton's interpretation even for equal speeds.

As the missile speed increases, the target's flight-path arc is reduced by an even greater factor. If $k = 3$, for example — the missile is three times faster than the target — then

$$\hat{t} = \frac{R}{V} \sin^{-1}\left(\frac{1}{3}\right) = \frac{1}{V} \cdot \frac{2\pi R}{2\pi / \sin^{-1}(\frac{1}{3})} = \frac{1}{V} \cdot \frac{2\pi R}{18.488825\ldots},$$

which represents an arc of only $19.5°$ that the target moves through during the interval $t = 0$ to $t = \hat{t}$.

There remains at least one more interesting question to answer about Houghton's problem: what is the pursuit path flown by the missile? The answer has a pleasing symmetry with the original problem — as it chases after its target, the missile flies along a circular path, too. This is not hard to prove with what we have already shown. Again, with reference to figure 2.4.1, the missile's position at time t is completely specified by the following two parametric time functions:

(2.4.7)
$$\theta(t) = \frac{V}{R}t$$

and

(2.4.8)
$$r(t) = kR\sin\left(\frac{V}{R}t\right).$$

What I'll now demonstrate is that these two expressions describe a circle centered on the point $(0, kR/2)$, with radius $kR/2$. (Notice that this result, once we've established it, tells us that the target and missile circles do not intersect if $k < 1$, which is simply the mathematics saying that there is no interception unless $k \geq 1$, that is, the missile is at least as fast as the target.) If our claim is correct, then if we write the equation of the missile's path in rectangular coordinates that equation should be

(2.4.9)
$$x^2 + \left(y - \frac{kR}{2}\right)^2 = \left(\frac{kR}{2}\right)^2.$$

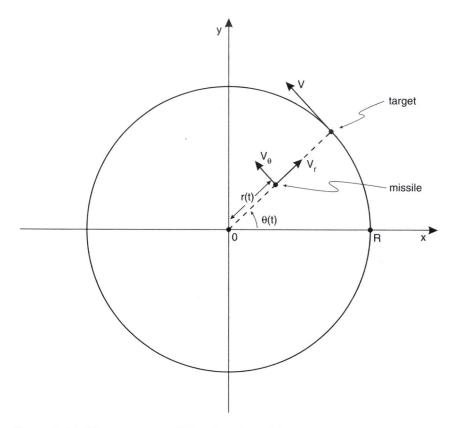

Figure 2.4.1 The geometry of Houghton's problem

To show that (2.4.9) is indeed satisfied by (2.4.7) and (2.4.8), notice that figure 2.4.1 shows that we can write

$$\frac{y}{r(t)} = \frac{y}{kR\sin(Vt/R)} = \sin(\theta) = \sin\left(\frac{V}{R}t\right)$$

and

$$\frac{x}{r(t)} = \frac{x}{kR\sin(Vt/R)} = \cos(\theta) = \cos\left(\frac{V}{R}t\right)$$

or

(2.4.10)
$$y = kR\sin^2\left(\frac{V}{R}t\right)$$

and

(2.4.11)
$$x = kR\cos\left(\frac{V}{R}t\right)\sin\left(\frac{V}{R}t\right).$$

Substituting (2.4.10) and (2.4.11) into the left-hand side of (2.4.9), we have

$$(kR)^2\cos^2\left(\frac{V}{R}t\right)\sin^2\left(\frac{V}{R}t\right) + \left[kR\sin^2\left(\frac{V}{R}t\right) - \frac{kR}{2}\right]^2$$

$$= (kR)^2\left\{\cos^2\left(\frac{V}{R}t\right)\sin^2\left(\frac{V}{R}t\right) + \left[\sin^2\left(\frac{V}{R}t\right) - \frac{1}{2}\right]^2\right\}$$

$$= (kR)^2\left\{\cos^2\left(\frac{V}{R}t\right)\sin^2\left(\frac{V}{R}t\right) + \sin^4\left(\frac{V}{R}t\right) - \sin^2\left(\frac{V}{R}t\right) + \frac{1}{4}\right\}$$

$$= (kR)^2\left\{\left[1 - \sin^2\left(\frac{V}{R}t\right)\right]\sin^2\left(\frac{V}{R}t\right) + \sin^4\left(\frac{V}{R}t\right) - \sin^2\left(\frac{V}{R}t\right) + \frac{1}{4}\right\}$$

$$= \frac{1}{4}(kR)^2$$

$$= \left(\frac{kR}{2}\right)^2,$$

which is, indeed, the *right*-hand side of (2.4.9) and we are done.

2.5 Pursuit of Invisible Targets

In this section I'll start by showing you a pursuit-and-evasion problem that most people think (incorrectly), upon first sight, to describe an impossible task. For that reason alone (besides being a *mathematically interesting* problem), it has become a popular question for textbook authors to include in their end-of-chapter problem sets, and here is how it appears in one such book:[8]

A coast-guard boat is hunting a rum runner in a fog. The fog rises disclosing the rum runner 4 miles distant and [then] immediately descends. The speed of the boat is three times that of the rum runner; and it is known that the latter will immediately depart at full speed on a straight course of unknown direction. What course should the boat take in order to overtake the rum runner?

This strikes most people as a "shaggy-dog" problem — if the coast guard don't even know in which direction the rum runner takes off *unseen into the fog*, how can they possibly catch him *even with* a faster boat? Notice that there is no preferential direction in which the rum runner will attempt his "invisible escape" — he might, if particularly nervy, try to be clever about his evasive action and head almost right *toward* the location of the coast-guard boat at the instant the fog lifted! In a thick fog, and with a quiet engine, he might even get away with that strategy. Picking a pursuit course at random might conceivably, by sheer luck, allow the coast-guard boat to run down the rum runner, but that approach is just a shot-in-the-dark, with no guarantee of success. It seems absurd, therefore, to think that the coast guard could have a strategy that will *ensure* catching the rum runner. Well, it *is* possible, and here's how the coast-guard boat can do it.

To set the problem up mathematically, the following nearly obvious observation is the crucial insight. If we denote as the origin of our coordinate system the location of the rum runner at the instant the fog momentarily lifts (which we'll define as time $t = 0$), then to *catch* the rum runner means that at the time of capture the rum runner and the coast-guard boat are the *same* distance from the origin. To be general in our analysis, let's write k_1 for the speed of the rum runner and k_2 for the speed of the coast-guard boat. Then, at time t the rum runner will be distance $k_1 t$, *in some unknown direction*, from the origin. Therefore, whatever the pursuit strategy may be for the coast-guard boat, it must also be distance $k_1 t$ from the origin if t is the time of interception. But in *what* direction?, you may again ask. The clever answer to that is that the coast-guard boat will try *all possible* directions — but *not* at the same time, of course! Here's how that can be done.

First, the coast-guard boat stays on the heading it had at the instant it spotted the rum runner and heads directly toward the point where the rum runner *would be if* it had headed directly toward the coast-guard boat. (We'll take this initial heading of the coast-guard boat to define the x-axis of our coordinate system, and so the boat travels along the x-axis *toward* the origin.) If the rum runner is actually at that point the chase is over, but that is very unlikely to be the case. More likely is that the rum runner is not there, and so the coast-guard boat switches to the second phase of its pursuit, and begins to travel along an *increasing spiral* path around the origin. The spiral should increase the coast-guard boat's distance from the origin at the same rate at which the rum runner's distance from the origin is increasing. The rum runner and the coast-guard boat are then *certain* to arrive at some common point at the same time.[9] And that certain encounter is *certain* to occur even before the coast-guard boat has made even one complete 360° swing around the origin.

So, to start, if d is the initial separation of the rum runner and the coast-guard boat, then, for the two to meet under the *assumption* that the rum runner makes straight for the coast-guard boat, we must have $k_1 T + k_2 T = d$, where T is the time it takes for the two to meet. That is, the coast-guard boat first travels, on its initial heading, for a distance

$$k_2 T = k_2 \frac{d}{k_1 + k_2} = d \frac{k_2/k_1}{1 + k_2/k_1}.$$

This eventually puts the coast-guard boat on the x-axis at a distance from the origin of

$$d - d \frac{k_2/k_1}{1 + k_2/k_1} = \frac{d}{1 + k_2/k_1},$$

and the coast-guard boat reaches this point at time

$$t = T = d/(k_1 + k_2).$$

Then, once at this distance from the origin, the coast-guard boat next begins to start swinging around the origin along an *ever expanding spiral* path; that is, the position vector of the boat is written as

$$(2.5.1) \qquad \mathbf{p}(t) = \frac{d}{1 + k_2/k_1} \cdot \frac{t}{T} e^{i\theta(t)}, \ t \geq T,$$

where $\theta(t)$ — the angle the position vector $\mathbf{p}(t)$ makes with the x-axis — is some function of time yet to be determined. $\theta(t)$ will be determined, you'll soon see, by the requirement that $\mathbf{p}(t)$ must be consistent with a speed of k_2 for the coast-guard boat. All we demand, at this point, is that $\theta = 0$ for $0 \leq t \leq T$. In fact, (2.5.1) can be simplified by inserting the expression for T:

$$(2.5.2) \qquad \mathbf{p}(t) = \frac{d}{1 + k_2/k_1} \cdot \frac{t}{d/(k_1 + k_2)} e^{i\theta(t)} = k_1 t e^{i\theta(t)},$$

where it is understood that the coast-guard boat is actually on this course only for $t \geq T$. At time $t = T$ the boat is at distance $k_1 T$ from the origin (as is the rum runner), and for all $t > T$ the rum runner and the boat are *always* the same distance, $k_1 t$, from the origin; $|\mathbf{p}(t)| = |k_1 t e^{i\theta(t)}| = k_1 t |e^{i\theta(t)}| = k_1 t$.

Now, the velocity vector of the coast-guard boat is

$$\frac{d\mathbf{p}(t)}{dt} = k_1 e^{i\theta(t)} + i k_1 t e^{i\theta(t)} \frac{d\theta}{dt}, \quad t \geq T,$$

or, using Euler's identity to expand the complex exponentials and then collecting real and imaginary terms,

$$(2.5.3) \qquad \frac{d\mathbf{p}(t)}{dt} = k_1 \left[\cos(\theta) - t \sin(\theta) \frac{d\theta}{dt} + i \left\{ \sin(\theta) + t \cos(\theta) \frac{d\theta}{dt} \right\} \right].$$

Since the *magnitude* of the velocity vector is the *speed* of the coast-guard boat, then

$$\left| \frac{d\mathbf{p}(t)}{dt} \right|^2 = k_2^2,$$

and so

$$k_2^2 = k_1^2 \left[\cos^2(\theta) - 2t \cos(\theta) \sin(\theta) \frac{d\theta}{dt} + t^2 \sin^2(\theta) \left(\frac{d\theta}{dt} \right)^2 \right.$$

$$\left. + \sin^2(\theta) + 2t \cos(\theta) \sin(\theta) \frac{d\theta}{dt} + t^2 \cos^2(\theta) \left(\frac{d\theta}{dt} \right)^2 \right],$$

or

(2.5.4)
$$k_2^2 = k_1^2 \left[1 + t^2 \left(\frac{d\theta}{dt} \right)^2 \right].$$

This is easily solved for $d\theta/dt$ to give

(2.5.5)
$$\frac{d\theta}{dt} = \frac{\sqrt{(k_2/k_1)^2 - 1}}{t},$$

which integrates immediately to

(2.5.6)
$$\theta(t) = \sqrt{\left(\frac{k_2}{k_1} \right)^2 - 1} \ln(t) + C, \ t \geq T,$$

where C is the constant of indefinite integration. Notice, carefully, that I have used the *positive* square root to swing the coast-guard boat around the origin in the *counterclockwise* sense, but I could just as well have used the negative root to swing the boat in the clockwise sense. It's an arbitrary choice. Since $\theta(T) = 0$, then if we write $\lambda = \sqrt{(k_2/k_1)^2 - 1}$ we can write $\lambda \ln(T) + C = 0$ $(C = -\lambda \ln(T))$ and so $\theta(t) = \lambda \ln(t) - \lambda \ln(T) = \lambda \ln(t/T)$ which says

(2.5.7)
$$t = Te^{\theta/\lambda}.$$

Since the distance of the coast-guard boat from the origin is $r = k_1 t$ for $t \geq T$, we have

$$r = k_1 Te^{\theta/\lambda},$$

or, upon substituting the expressions for T and λ, we have the polar equation[10] for the second part ($t \geq T$) of the coast-guard boat's path:

$$r = k_1 \cdot \frac{d}{k_1 + k_2} \cdot \exp \left(\theta \Big/ \sqrt{\left(\frac{k_2}{k_1} \right)^2 - 1} \right)$$

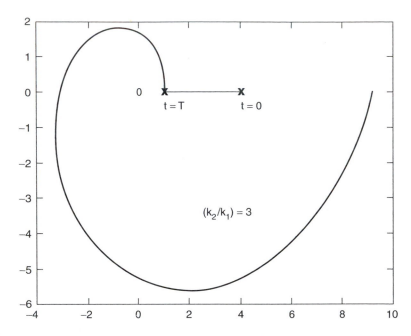

Figure 2.5.1 The pursuit path of the coast-guard boat

or,

(2.5.8)
$$r = d \, \frac{\exp\left(\theta / \sqrt{(k_2/k_1)^2 - 1}\right)}{1 + (k_2/k_1)}.$$

In particular, for the problem that opened this section, with $d = 4$ and $(k_2/k_1) = 3$, we have $r = e^{\theta/\sqrt{8}}$. That path for the coast-guard boat (the two crosses mark the $t = 0$ and $t = T$ points, and the O is the rum runner's position when sighted) is shown in figure 2.5.1.

This increasing spiral path strategy suggests a solution to a different sort of search for (pursuit of) an invisible target; this variant was first proposed by the American mathematician Richard Bellman (1920–1984), in a 1956 paper in the *Bulletin of the American Mathematical Society*. To understand Bellman's problem, consider the following not at all unlikely situation. A man gets into his boat somewhere along a long, straight beach, and rows out to sea for a while. Then he stops rowing and begins to fish while his boat continues to drift randomly with the waves and tide. He becomes so involved with fishing that he fails to notice a slowly developing fog until, suddenly, he realizes he can no longer see the shore and, in fact, that he has completely lost his orientation. He doesn't know in which direction the shore lies, or even how distant it is. He is really lost!

Or is he? Actually not, because if he simply begins to row along an expanding spiral path it should now be clear to you that, *eventually*, he is *certain* to encounter the shore. Depending on how fast the spiral expands, and in which direction he initially begins to row, the man may have to row for a "long" time, yes, but in the end he *will* hit land. He isn't really "lost" at all.

Suppose now we add one more detail to the fisherman's predicament. As before, he doesn't know the direction to shore, but now he *does* know the *distance* from shore. In this case we can say some very specific things. Without any loss of generality, let's take the distance to shore as our unit distance. A path that ensures that the man eventually reaches shore doesn't have to be anything as sophisticated as an expanding spiral in this case; first he rows, in *any* direction he wishes, unit distance. Assuming that doesn't get him to shore immediately (it *might*, if he is *really* lucky in his initial direction guess!), he then starts rowing along a circular path with radius one centered on his original position. By the time he has rowed all the way around the circle, *at most*, he is *certain* to have encountered the shore. That is, *at most* he rows a distance of $1 + 2\pi = 7.283$. This calculation led Bellman to ask if this is the *smallest* possible maximum rowing distance? Or, as it is technically called, what is the *minimax path* for the lost fisherman?

It didn't take long for the answer to appear — in 1957 the American mathematician John Isbell (1930–), then at the Institute for Advanced Study in Princeton, showed that a significantly shorter maximum length path exists — see Isbell (1957). As Isbell opens his paper (which appeared in a relatively obscure publication),

Given a base point p and a line L at unit distance from p in a plane [find] a [curve C] of minimum length such that C joins p to L, and if C is rotated about p through any angle, it still meets L. In other words, one is at known distance from the line L but does not know its direction.

Isbell notes that "no systematic method for [this and similar] problems is available," but he does provide a general and powerful insight into discovering a solution to Bellman's problem: "the path C [must] meet every tangent to the unit circle about p." With reference to figure 2.5.2, then, here is Isbell's solution in his own words:

> Being at p and unoriented, imagine a clock face. Walk toward one o'clock for $\sqrt{4/3}$ units. (This takes you to a vertex of a circumscribed regular hexagon.) Then turn on the tangent which strikes the unit circle at two o'clock. Follow the circle to nine o'clock and continue on a tangent. Upon striking the line which is tangent to the unit circle at twelve o'clock you have swept through all tangents to the unit circle, and that in a path of minimum length [a proof of which Isbell then outlines, but which I'll not include here].

By elementary geometry, the described curve has length

$$\sqrt{\frac{4}{3}} + \frac{1}{2}\sqrt{\frac{4}{3}} + \frac{9-2}{12}2\pi + 1 = \sqrt{3} + \frac{7\pi}{6} + 1 = 6.397,$$

which is *significantly* less (more than 12% less) than $1 + 2\pi$.

A few years later Isbell's solution was generalized a bit, in Gluss (1961). There the search is for a circle rather than a line. Gluss wrote, "The problem appears to be of some practical significance, since it is equivalent to that of searching for an object a given distance away which will be spotted when we get sufficiently close — that is, within a specific radius." Gluss's solution contains Isbell's as a special case, because as the radius of the circle being searched for increases without limit that part of the circle in the neighborhood of p approaches Isbell's straight line L (and Gluss's minimax path approaches Isbell's path of figure 2.5.2).

To end this section, let me give you a challenge problem based on the earlier rum runner/coast-guard boat pursuit-and-evasion analysis. Suppose the *total* time, from first sighting to interception, for some particular values of k_2/k_1 and d, with the coast-guard boat 20% faster than the rum runner, is denoted by $\widehat{T}$. What would be the new total time, in terms of $\widehat{T}$, from first sighting to interception, if a new

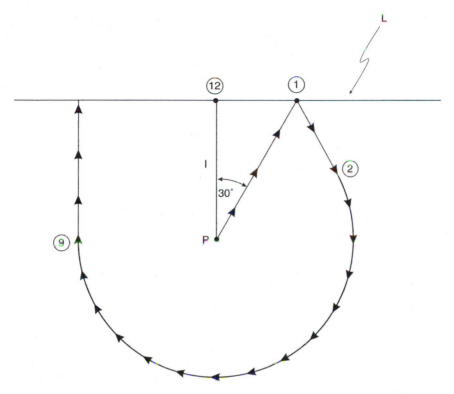

Figure 2.5.2 Isbell's solution curve to Bellman's problem

coast-guard boat is purchased that is 40% faster than the rum runner (but d remains the same)? That is, by what factor is the total elapsed capture time reduced by the faster boat? Assume the worst-case scenario of the coast-guard boat having to make nearly one complete swing around the origin. If you have trouble with the calculations you can take a look at my solution in appendix F.

2.6 Proportional Navigation

As shown earlier in this chapter, if a missile uses pure pursuit to intercept a target, and if the missile flies faster than twice the speed of the target, then the missile will have to be able to generate an infinite acceleration at the end of the attack. That, of course, is not a physically

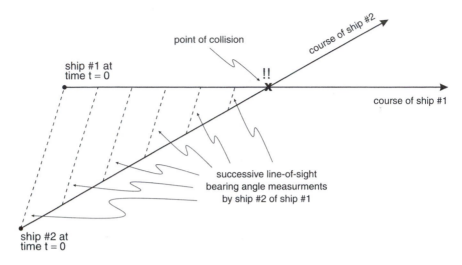

Figure 2.6.1 A *constant* bearing angle means *collision!*

reasonable expectation. For this reason alone, it becomes interesting to explore strategies for pursuit missiles to employ other than pure pursuit. One such alternative strategy that has long been popular is the so-called *proportional navigation* strategy. Surprisingly, perhaps, the basic idea behind proportional navigation actually predates modern intercept missiles by *centuries*.

Imagine two ships at sea, as shown in figure 2.6.1, each on a straight line course that crosses the course of the other. The fact that the *courses* cross does *not* mean the two ships will collide — the mere crossing of the courses is a necessary condition for a collision, obviously, but it isn't a sufficient condition since that additionally requires the two ships be at the course crossing point at the *same* time. Suppose now that the captain of one of the ships periodically takes sightings of the other ship by measuring the so-called *bearing angle*. In other words, he measures at a number of different times the angle from some fixed reference line, such as the line through his own ship from stern to bow, to the other ship as indicated by the line-of-sight (LOS) dashed lines in figure 2.6.1. As the earliest sailors learned (or, at least as those who survived to go to sea again learned!) a *constant* bearing angle as the two ships approach the crossing point of their courses means collision. Alternatively, a *rotating* LOS means *no* collision. Of course, what is bad

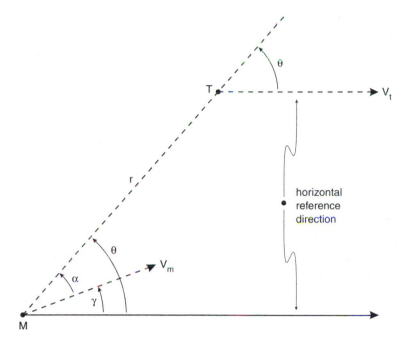

Figure 2.6.2 The geometry of proportional navigation

for a sailor is good for a missile whose whole reason for existing is to cease to exist in a spectacular interception and explosion. Such a missile *seeks* to achieve a constant LOS bearing angle trajectory as it pursues its target.

An intercept missile could, then, constantly measure the LOS bearing angle to its target, using for example, the IR seeker technology described in section 2.3 — if there is zero rotation of the LOS then no steering commands to the missile's air-flow surfaces are generated by the flight control guidance system, while any rotation of the LOS does generate steering commands such that the LOS rotation rate is *reduced*. Figure 2.6.2 shows the geometry of an interceptor missile pursuing a nonmanuevering target, where the direction of the target's straight-line, constant speed (V_t) motion defines the horizontal reference direction. From the figure we have $\theta = \alpha + \gamma$, where θ is the LOS angle measured with respect to the horizontal reference direction and γ is the instantaneous angle the missile's velocity vector makes with the reference direction. We'll actually be more interested in $d\gamma/dt$, the turn-*rate* of the missile, than in γ itself. Indeed, by definition, the guidance system of the missile is one of "proportional navigation" if the turn-rate of the missile is *proportional* to the LOS *rotation rate*, that is, if

$$(2.6.1) \qquad \frac{d\gamma}{dt} = N\frac{d\theta}{dt},$$

where N is the so-called *navigation constant* (typically a number somewhere between 3 and 5 — you'll see why at the end of the analysis). This equation tells us, for example, that if the target veers *away* from the missile (if the LOS angle θ *increases*) then the missile velocity vector angle γ also increases, that is, the missile turns *toward* the target just as we would expect a missile that is chasing after a target to do. As the chase evolves in time the missile tries to swing itself into a *constant* LOS angle collision course, called a *navigational course*. An observer on the target would see the inbound missile *not* pointed at the target, but rather tilted off from a direct head-on attack by what is called the *lead angle*.

Since $\theta = \alpha + \gamma$, then

$$(2.6.2) \qquad \frac{d\alpha}{dt} = \frac{d\theta}{dt} - \frac{d\gamma}{dt},$$

or, using (2.6.1),

$$(2.6.3) \qquad \frac{d\alpha}{dt} = \frac{d\theta}{dt} - N\frac{d\theta}{dt} = (1 - N)\frac{d\theta}{dt}.$$

Integrating (2.6.3), with the initial conditions $\alpha(t = 0) = \alpha_0$ and $\theta(t = 0) = \theta_0$, we have

$$(2.6.4) \qquad \alpha = (1 - N)\theta + \alpha_0 + (N - 1)\theta_0,$$

or, making the definitions

$$(2.6.5) \qquad k = N - 1, \phi_0 = \alpha_0 + (N - 1)\theta_0,$$

(2.6.4) becomes

$$(2.6.6) \qquad \alpha = \phi_0 - k\theta.$$

With r as the radial distance from the target (which is at our coordinate system origin) to the missile, as shown in figure 2.6.2, we can write the radial and transverse components (with respect to the LOS) of the closing velocity between the target and the missile as

$$(2.6.7) \qquad \frac{dr}{dt} = V_t \cos(\theta) - V_m \cos(\alpha)$$

and

$$(2.6.8) \qquad r\frac{d\theta}{dt} = V_m \sin(\alpha) - V_t \sin(\theta).$$

Using (2.6.6) to eliminate α, we arrive at the differential equations for proportional navigation:

$$(2.6.9) \qquad \frac{dr}{dt} = V_t \cos(\theta) - V_m \cos(\phi_0 - k\theta) = V_r(\theta)$$

and

$$(2.6.10) \qquad r\frac{d\theta}{dt} = V_m \sin(\phi_0 - k\theta) - V_t \sin(\theta) = V_\theta(\theta),$$

where (you'll recall) $k = N - 1$ with N an adjustable value called the navigation constant. As we saw earlier in Hathaway's dog-and-duck problem, however, simply having differential equations doesn't mean we can solve them, and the proportional navigation equations of (2.6.9) and (2.6.10) are no exception.

We can *start* a solution as follows. Dividing (2.6.9) by (2.6.10) eliminates explicit time, that is,

$$(2.6.11) \qquad \frac{dr}{r} = \frac{V_r(\theta)}{V_\theta(\theta)} d\theta,$$

which can be formally integrated to give the LOS distance between the missile and the target:

$$r = r_0 \exp\left(\int_{\theta_0}^{\theta} \frac{V_r(u)}{V_\theta(u)} du \right).$$

The integral in the exponent is impossible to actually evaluate, unfortunately, except in the two very special (and *not* of any practical interest) cases $k = 0$ and $k = 1$, when the navigation constant is either $N = 1$ or $N = 2$ — see Shukla and Mahapatra (1990) for the grubby mathematical details. So, alas, it appears we are stopped — but, actually, not entirely. As shown in Guelman (1971) we *can* say something, in general, about the LOS rotation rate $d\theta/dt$, which plays a crucial role in the "buildability" of a proportional navigation missile, much as does the acceleration for a pure pursuit missile. It is clear that, whatever $d\theta/dt$ is, it must be *finite* if a constructable IR seeker is to be able to measure it, that is, for the seeker to be able to rotate as fast as the LOS angle is changing. And, perhaps amazingly, even without solving the differential equations of proportional navigation pursuit we can place an upper *finite* bound on the LOS rotation rate, a bound that is a function only of the missile and target speeds (V_m and V_t) and the navigation constant (N).

The derivation of the upper bound on the LOS rotation rate consists of two steps. First, I'll establish an abstract, symbolic, preliminary result. And second, this abstract result will then be translated into a form expressed in terms of the actual missile and target parameters (V_m, V_t, and N). So, first, Step 1:

Theorem: if $dV_\theta/d\theta < V_r(\theta)$ — see (2.6.9) and (2.6.10) again for what $V_r(\theta)$ and $V_\theta(\theta)$ are — then $d\theta/dt$ is a monotonic *decreasing* function of time.

Proof: since the inequality $dV_\theta/d\theta < V_r(\theta)$ is given, we can next rewrite it as

$$\frac{dV_\theta/dt}{d\theta/dt} < V_r,$$

or

$$\frac{dV_\theta/dt}{r\,d\theta/dt} < \frac{V_r}{r}.$$

Recalling (2.6.10) for $r\,d\theta/dt$, and (2.6.9) for V_r, we have

$$\frac{dV_\theta/dt}{V_\theta} < \frac{dr/dt}{r},$$

or

$$\frac{dV_\theta}{V_\theta} < \frac{dr}{r},$$

or, integrating over arbitrary intervals of θ and r, that is, over $\theta_i < \theta < \theta_f$ and $r_i < r < r_f$ as time varies over the interval $t_i < t < t_f$, we have

$$\frac{V_\theta(\theta_f)}{V_\theta(\theta_i)} < \frac{r_f}{r_i}.$$

Using (2.6.10) one last time, we have $r_i (d\theta/dt)|_{\theta=\theta_i} = V_\theta(\theta_i)$ and $r_f (d\theta/dt)|_{\theta=\theta_f} = V_\theta(\theta_f)$, and so

$$\frac{r_f (d\theta/dt)|_{\theta=\theta_f}}{r_i (d\theta/dt)|_{\theta=\theta_i}} < \frac{r_f}{r_i},$$

or, at last,

$$\boxed{\left(\frac{d\theta}{dt}\right)|_{\theta=\theta_f} < \left(\frac{d\theta}{dt}\right)|_{\theta=\theta_i},}$$

which establishes the theorem and completes Step 1. That is, as long as $\theta_f > \theta_i$, that is, $t_f > t_i$, then the LOS rotation rate when $t = t_f$ will be *less* than it was when $t = t_i$.

Step 2 of our derivation of a finite upper bound on the LOS rotation rate is establishing a *practical* sufficient condition (one in terms of physically observable parameters) that ensures the truth of our abstract inequality $dV_\theta/d\theta < V_r(\theta)$. To do that, let's *assume* $dV_\theta/d\theta < V_r(\theta)$ is true, and see what that assumption implies. Certainly it must be true, assuming the missile actually does intercept the target (which, to state the obvious, means it's getting *closer* to the target), that we can write $dr/dt < 0$, that is, from (2.6.9) we have $V_r(\theta) < 0$. So, multiplying through the inequality $dV_\theta/d\theta < V_r(\theta)$ by the *negative* quantity $V_r(\theta)$, we *reverse* the sense of the inequality and arrive at

(2.6.12) $$V_r \frac{dV_\theta}{d\theta} > V_r^2.$$

From (2.6.10) we have

$$\frac{dV_\theta}{d\theta} = -V_m k \cos(\phi_0 - k\theta) - V_t \cos(\theta)$$

and so, inserting this expression, as well as (2.6.9) and (2.6.10) into (2.6.12):

$$[V_t \cos(\theta) - V_m \cos(\phi_0 - k\theta)][-V_m k \cos(\phi_0 - k\theta) - V_t \cos(\theta)] >$$
$$[V_t \cos(\theta) - V_m \cos(\phi_0 - k\theta)]^2.$$

If this is expanded, and if we write $v = V_m/V_t$, then a little bit of algebra gives us (you should confirm this)

$$(k-1)v^2 \cos^2(\phi_0 - k\theta) - 2\cos^2(\theta) - v(k-3)\cos(\theta)\cos(\phi_0 - k\theta) > 0.$$

(2.6.13)

To go any further with (2.6.13) we need to make the following clever observation, due to Guelman (1991). If we rewrite the differential equation of proportional navigation (2.6.11) as

(2.6.14)
$$r\frac{d\theta}{dr} = \frac{V_\theta(\theta)}{V_r(\theta)},$$

then it is clear that any *constant* value of θ $(=\theta_c)$ that satisfies the conditions $V_\theta(\theta_c) = 0$, $V_r(\theta_c) \neq 0$, satisfies (2.6.14) and so is a solution. Since we are working under the assumption that the LOS rotation rate is constantly decreasing, then in fact θ *is* approaching a constant value. Thus, in the *terminal phase* of the missile trajectory we take $V_\theta = 0$, that is, we set (2.6.10) equal to zero to get[11]

$$V_m \sin(\phi_0 - k\theta) - V_t \sin(\theta) = 0,$$

or, dividing through by V_t,

(2.6.15)
$$v\sin(\phi_0 - k\theta) = \sin(\theta).$$

Since

$$v^2 \cos^2(\phi_0 - k\theta) + v^2 \sin^2(\phi_0 - k\theta) = v^2,$$

then

$$v^2 \cos^2(\phi_0 - k\theta) = v^2 - v^2 \sin^2(\phi_0 - k\theta)$$

$$= v^2 - v \sin(\phi_0 - k\theta)v \sin(\phi_0 - k\theta),$$

and using (2.6.15) gives us

(2.6.16) $$v^2 \cos^2(\phi_0 - k\theta) = v^2 - \sin(\theta)v \sin(\phi_0 - k\theta).$$

Also, since $\cos^2(\theta) = 1 - \sin^2(\theta) = 1 - \sin(\theta)\sin(\theta)$, then, using (2.6.15) once more, we have

(2.6.17) $$\cos^2(\theta) = 1 - v \sin(\theta) \sin(\phi_0 - k\theta).$$

Remember, (2.6.15), (2.6.16), and (2.6.17) are *not* true in general, but only in the *terminal* or *end-game* phase of the missile attack, where we are using the fact that θ is approaching a constant value.

If we now insert (2.6.16) and (2.6.17) into (2.6.13) then, again with just a bit of algebra (and again, I encourage you to check this), we arrive at

(2.6.18) $$(k - 1)v^2 - 2 - (k - 3)v \cos[\phi_0 - (k + 1)\theta] > 0.$$

Now, as θ varies then of course $-1 \leq \cos[\phi_0 - (k + 1)\theta] \leq 1$. So, the smallest the left-hand side (LHS) of (2.6.18) can be occurs when

$$\cos[\phi_0 - (k + 1)\theta] = +1 \text{ if } k > 3,$$

$$\cos[\phi_0 - (k + 1)\theta] = -1 \text{ if } k < 3.$$

For the $k > 3$ case we can factor the LHS of (2.6.18) when it is smallest as

$$(k-1)v^2 - (k-3)v - 2 = [(k-1)v + 2](v-1),$$

and for the $k < 3$ case we can factor the LHS of (2.6.18) when it is smallest as

$$(k-1)v^2 + (k-3)v - 2 = [(k-1)v - 2](v+1).$$

That is, we require that

$$[(k-1)v + 2](v-1) > 0 \text{ for } k > 3,$$
$$[(k-1)v - 2](v+1) > 0 \text{ for } k < 3.$$

If we assume that $v > 1$, then the $k > 3$ condition is satisfied, and the $k < 3$ condition will also be satisfied if $(k-1)v > 2$. Thus, for *all* k, we require $(k-1)v > 2$ to ensure that our starting assumption $dV_\theta/d\theta < V_r(\theta)$ is true (which, in turn, means that the LOS rotation rate is decreasing as the missile approaches its target — which means the missile is on a collision course with the target).

Recalling the definitions of k and v, we thus have

$$(N - 1 - 1)\frac{V_m}{V_t} > 2$$

or, at last, we have a sufficient, *practical* condition that *ensures* a decreasing LOS rotation rate for a proportional navigation missile pursuing a nonmanuevering target for the usual case of $V_m > V_t$:

$$\frac{N-2}{2}V_m > V_t.$$

You can see that this inequality is satisfied for *any* $N > 4$.

Today, the theory discussed in this chapter is mostly applicable to air-to-air and ground-to-air missiles launched at very long range from their targets, as great as 100 miles or more; despite what Hollywood

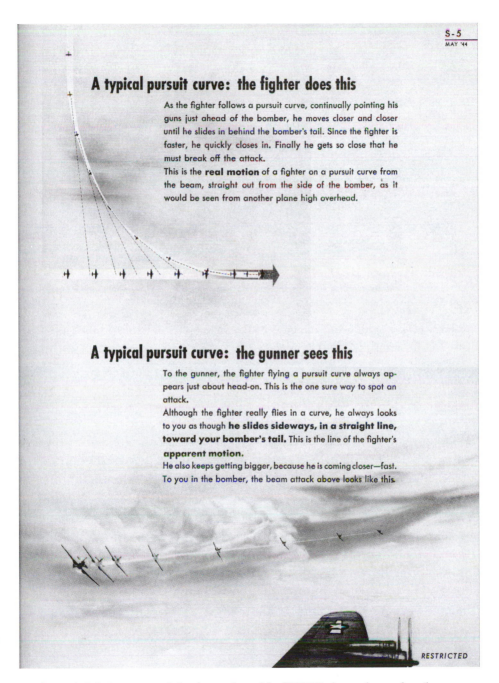

Figure 2.6.3 Pure pursuit in the real world of WWII air-to-air combat (image courtesy of the Hagley Museum and Library, Wilmington, Delaware)

Figure 2.6.4 The author next to the ball turret of a B-17G bomber. The men who flew into mortal combat against high-speed pursuit fighters in machines such as this were almost unbelievably brave.

showed in the 1986 movie *Top Gun*, modern combat pilots rarely engage in gun battles with other aircraft. This is very different from World War II, when aircraft fought at ranges no greater than their machine guns could be accurately fired (several hundred to at most a couple of thousand feet or so). In those days, pilots and gunners were routinely trained in the operational aspects of pure pursuit; those were the days when air-to-air combat was up-close-in-your-face-and-personal! Figure 2.6.3, for example, is taken from a 1944 training manual produced by the Sperry Gyroscope Company, used in showing bomber gunners how to recognize when their aircraft was under a pure pursuit fighter attack.

The Sperry Gyroscope Company was famous for its airborne ball turret machine gun battle station, used on both the B-17 Flying Fortress and B-24 Liberator bombers. The gunners in those turrets faced pure pursuit fighter attacks on nearly every mission they flew, and it was a deadly business, indeed. One of the most famous poems of the war was Randall Jarrell's "The Death of the Ball Turret Gunner" (1945), and anyone who is fascinated by the mathematics of air-to-air combat should also keep the human toll in mind as well, a toll powerfully expressed in Jarrell's poem:

> From my mother's sleep I fell into the State,
> And I hunched in its belly till my wet fur froze.
> Six miles from earth, loosed from my dreams of life,
> I woke to black flak and the nightmare fighters.
> When I died they washed me out of the turret with a hose.[12]

Jarrell (1914–1965) served in the war as a control tower operator in the Army Air Corps, and knew first-hand of what he wrote.

As a final comment on figure 2.6.3, the characteristic visual signature of a pure pursuit fighter attack was common knowledge among many writers of WWII war fiction, even if some of them didn't really understand it. For example, consider the following passage from John Hersey's 1959 novel *The War Lover*, told in narrative form by the copilot of a B-17, who observes attacking German fighters while on a bombing mission over the Focke-Wulf aircraft factory at Bremen (p. 85):

> I saw some enemy fighters this time, but it was hard to take them seriously, because all I could get was a feeling that they were *skidding*. They weren't flying, they were sliding sideways. The impression was a product of our speed plus or minus theirs. When they came at us head-on, the rate of closure was six hundred miles per hour, but it was never a bee line; they *skidded*.

As the Sperry illustration in figure 2.6.3 makes clear, however, if the attacking fighters appeared to be *skidding* to the narrator then *his* plane was the object of the fighters' attention — he would have done well to have taken them most seriously!

Chapter 3

Cyclic Pursuit

3.1 A Brief History of the *n*-Bug Problem, and Why It Is of *Practical* Interest

In the May 1877 issue of *Nouvelle correspondance mathématique* the French mathematician Edouard Lucas (1842-1891) *formally* posed the following pursuit problem, the first really new innovation in pursuit questions since Bouguer's original problem:

> Three dogs are placed at the vertices of an equilateral triangle; they run one after the other. What is the curve described by each of them?

The answer was not long in coming: in the August issue Henri Brocard (mentioned in the previous chapter) stated that, if we suppose the dogs to start at the same time and to run with the same speed, then the pursuit curve for each dog is a logarithmic spiral. In the next section we'll go through the mathematics of Brocard's answer, but for now I'll limit myself to outlining how the problem itself evolved over time.

But even before I do that, if you read the opening to the previous paragraph again perhaps you'll wonder why I emphasized the word *formally* when introducing Lucas's problem. The reason is that the problem had actually appeared *years earlier* on the famous Cambridge University Mathematical Tripos Examination of January 1871. That early appearance of cyclic pursuit is due to one R. K. Miller — I have been unable to learn anything about Miller, with only his name appearing in a brief Letter-to-the-Editor of *Mathematical Questions and Solutions* (50) 1889, p. 51 — who apparently anticipated Lucas by six years. However, since Lucas was the first to publish the problem statement in a mathematics journal, by convention he gets the *formal* credit.

The problem surfaced again in a rather obscure journal in Hackett (1908), which presents a very modern *numerical analysis* approach to studying the differential equations of cyclic pursuit. Hackett says not a word about Miller, Lucas, or Brocard, but he also does *not* impose the symmetrical constraint of the three dogs being initially at the vertices of an equilateral triangle (neither, for that matter, did Miller). Ten years later, in a textbook, the English-born and–trained American mathematician Harry Bateman (1882–1946) wrote[1] of the nonsymmetrical three-dog cyclic pursuit (with the dogs replaced by boys). In a footnote he cites Hackett, but attributes the problem to "Prof. Morley" (Frank Morley, Head of the Mathematics Department at The Johns Hopkins University — where Bateman taught from 1912 to 1917 — who you'll recall appeared in the previous chapter as the originator of a special pursuit curve called the *curve of ambience*). Bateman's footnote indicates that Hackett learned of the *non*-equilateral triangle pursuit problem from Morley. Morley was particularly interested in the question of mutual capture — would all three dogs/boys *simultaneously* meet, or would *first* two dogs/boys meet followed by a *later* meeting of the now combined two dogs/boys with the third dog/boy? Another question about nonsymmetrical cyclic pursuit that interested Morley was what the speeds of each of the three dogs/boys should be so as to have their instantaneous triangle always be similar to their initial triangle. Curiously, Miller had answered that second question decades before, on the Tripos Exam, where he gave the answers in the problem statement, asking only that the students verify them.

The problem of cyclic pursuit remained one known only to mathematicians through the first half of the 20th century, until the Polish mathematician Hugo Steinhaus (1887–1972) included it in his 1950 puzzle book *Mathematical Snapshots*. But what really brought it widespread fame was its use by Martin Gardner in his famous "Mathematical Games" column in *Scientific American* magazine (July 1965). As presented there, the problem became one of four bugs initially at the corners of a square. (This particular version of cyclic pursuit actually predates Gardner — the four-bugs-on-a-square problem had appeared far less famously several years earlier, in a book by L. A. Graham, *Ingenious Mathematical Problems and Methods*, Dover 1959, p. 40, with a correct solution given on pp. 203–205.) At time $t = 0$ each bug begins crawling toward its nearest neighbor, all moving at the same (constant) speed. The questions Gardner posed to his readers were (a) what path is traveled by each bug? and (b) how far does each bug travel until they all meet (it is clear, by symmetry, that all four bugs will *simultaneously* arrive at the center of the original square)?

The editors of *Scientific American* evidently immediately realized the popular appeal of the problem — they devoted the magazine's cover art to an illustration of what Aravind (1994) called the bugs' "square dance" as they spiral into the center of the square. From that point on the problem became known as the "four-bug cyclic pursuit" problem, which is most obviously generalized — first, apparently, even before Gardner's column appeared, by Clapham (1962) — to n bugs initially on the vertices of a regular n-gon. I'll show you the solution to this and other generalizations of the problem later in this chapter.

Now, before plunging into the mathematics of cyclic pursuit, let me tell you just a bit about why the problem is still receiving a lot of continuing attention from modern mathematicians. The problem is, of course, intrinsically interesting from a pure mathematical point of view. But it also promises to give insight into some naturally occurring phenomena that are commonly seen and yet remain puzzling even to experts. I'll give you just two examples of this, both from wildlife behavior.

In a provocatively titled paper, Bruckstein (1993) quotes from physicist Richard Feynman's famous book *Surely You're Joking, Mr. Feynman!*, in which Feynman is pondering how ants leave some sort of trail on the ground for other ants to follow to locations of food:

> One question that I wondered about was *why the ant trails look so straight and nice*. The ants look as if they know what they are doing, as if they have a good sense of geometry. Yet the experiments that I did to try to demonstrate their sense of geometry didn't work.

Bruckstein then goes on to provide his readers with the "solution" that Feynman finally arrived at:

> After some further experimentation, Feynman did find an explanation: The ants try to follow the original randomly found and marked path, but they coast on the wiggly path and leave it here and there, some to find the trail again. Each ant straightens the trail a bit, and their collective effort makes up for their lack of any sense of geometry.

As Bruckstein put it, using the imagery of pursuit, "If a 'pioneer' ant shows the way to the food along a random path it marks, and other ants follow in a row, *each ant pursuing the one in front of it* [my emphasis], their path becomes a straight line connecting the anthill and the food location." Bruckstein's paper then develops that claim mathematically. Bruckstein ends his paper with a comment that leads me to my second example: "The analysis of *global* [my emphasis] behavior that results from simple and *local* [my emphasis] interaction rules is a fascinating subject of investigation and may even lead to a better understanding of natural and artificial animal colony behavior. Such ideas could also be of use in problems arising in robotics, for example."

The concept expressed by Bruckstein has been proposed as an explanation for one of the most dramatic, beautiful, and downright *mysterious* sights in the sky — a large flock of birds flying in formation and *turning* as a single entity. From the front to the rear of the flock may be hundreds of feet, and yet the birds appear to be in instant communication, just as if they all had little radios on their heads with the front bird broadcasting "Okay, on the count of three we all turn left — one, two, three, TURN!" Well, we know they aren't wearing little radios, and so *how do* they achieve such amazing synchronization? The answer *may* be that each bird simply follows (pursues) the bird(s) immediately in front of it. This *local*, simple, cyclic pursuit rule for each individual bird may be all that is required to generate the amazing *global* flight behavior exhibited by the entire flock.

So, keep the ants and the birds in mind as we discuss bugs in the rest of this chapter. Cyclic pursuit mathematics is fascinating in its own right, yes, but it does appear to have application to matters far more significant than simply bugs chasing after other bugs.

3.2 The Symmetrical n-Bug Problem

The geometry of n bugs in a symmetric cyclic pursuit is shown in figure 3.2.1, for the situation of any two adjacent bugs, A and B. Without loss of generality I've chosen our coordinate axes such that bug A is initially on the horizontal axis (and so, of course, bug B is initially at angle $2\pi/n$). The initial distance between bugs A and B (between *any* two adjacent bugs, in fact) is then $2r(0)\sin(\pi/n)$, where the initial distance from the center C of the bugs' regular n-gon to each of the bugs is $r(0)$, and $r = r(t)$ denotes that distance for $t \geq 0$. Each bug always moves directly toward its nearest counterclockwise neighbor with the same speed v (I am following the notation and approach of Watton and Kydon (1969)), that is, the speed of A is v directly along the line joining A to B, as shown in the figure. We can resolve this speed into two components, a radial component v_r directed inward toward C, and a transverse component v_θ directed perpendicular to v_r. From figure 3.2.1, or from our results at the beginning of section 2.3, these speed components are

$$(3.2.1) \qquad v_r = \frac{dr}{dt} = -v\sin\left(\frac{\pi}{n}\right)$$

and

$$(3.2.2) \qquad v_\theta = r\frac{d\theta}{dt} = v\cos\left(\frac{\pi}{n}\right),$$

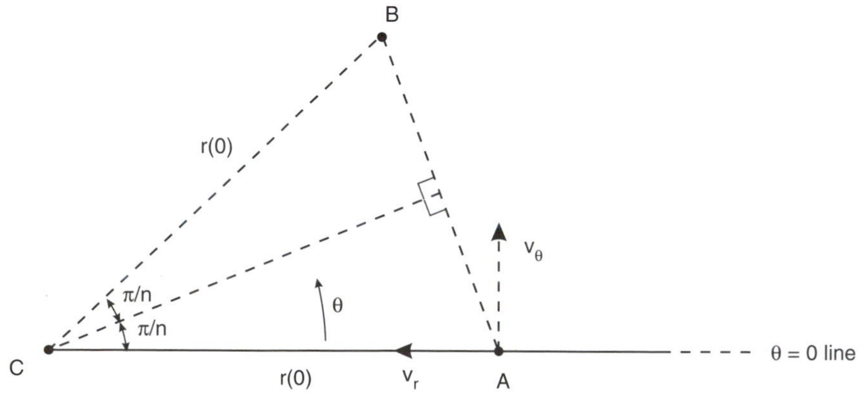

Figure 3.2.1 The geometry of the symmetrical n-bug problem

where the minus sign on the v_r equation, of course, simply reflects the physically obvious statement that v_r is directed *inward* toward C, that is, r *decreases* with *increasing* time. By symmetry, the n bugs will *always* be on the vertices of a regular n-gon, and so the central angle between any two adjacent bugs is always the *constant* $2\pi/n$.

At this point we can immediately integrate (3.2.1) and (3.2.2) to get expressions for $r(t)$ and $\theta(t)$, the polar coordinates for bug A at time t — and **you** (!) will be asked to do that, soon, in an upcoming challenge problem — but for now let's take a slightly different approach. From the chain rule in calculus we know that

$$\frac{d\theta}{dt} = \frac{d\theta}{dr} \cdot \frac{dr}{dt}$$

or, using (3.2.1) and (3.2.2),

$$\frac{v}{r} \cos\left(\frac{\pi}{n}\right) = \frac{d\theta}{dr} \cdot \left\{-v \sin\left(\frac{\pi}{n}\right)\right\},$$

which, upon cancelling the v on both sides (remember this cancellation!) reduces to

$$\frac{dr}{d\theta} = -r\frac{\sin(\pi/n)}{\cos(\pi/n)} = -r\tan\left(\frac{\pi}{n}\right),$$

or

(3.2.3) $$\frac{dr}{r} = -\tan\left(\frac{\pi}{n}\right)d\theta,$$

which immediately integrates to Brocard's logarithmic spiral. In polar coordinates A's path is given by

(3.2.4) $$r(\theta) = r(0)\exp\left(-\theta\tan\left(\frac{\pi}{n}\right)\right).$$

For Martin Gardner's four-bug problem ($n = 4$), mentioned in the previous section, (3.2.4) reduces to the particularly simple

$$r(\theta) = r(0)e^{-\theta}.$$

Before continuing with our analysis, I should emphasize one previous point that we sailed by pretty fast. You'll recall that in deriving (3.2.3), the differential equation for the path of a bug, the bugs' common speed v cancelled. That is, the value of v does *not* have to be anything in particular; we have been tacitly taking it as a *constant* (the same constant for each bug), but all that is really required is that v be the same for each bug — even a *function of time*, as long as it is the *same* function of time. Then the cancellation still goes through with no problem. This observation apparently escaped explicit mention in the literature until Watton and Kydon (1969).

Now, with that said, let's actually *do* take the bugs' speed to be the same constant. We can then easily calculate the total time T, from the start of the cyclic pursuit ($r = r(0)$) until the mutual collision of all n bugs at C ($r = 0$), as follows. Using (3.2.1),

$$T = \int_0^T dt = \int_{r(0)}^0 \frac{dr}{(dr/dt)} = \int_{r(0)}^0 \frac{dr}{-v\sin(\pi/n)},$$

or

(3.2.5)
$$T = \frac{1}{v}\int_0^{r(0)} \frac{dr}{\sin(\pi/n)} = \frac{r(0)}{v\sin(\pi/n)}.$$

Notice that as $n \to \infty$ we have $T \to \infty$. What this particular result means, physically, is that as $n \to \infty$ the regular n-gon "becomes" indistinguishable from a *circle*. When our infinity of bugs start their cyclic chase they all are traveling along that circle and so do *not* spiral inward to C, that is, there will *never* be a mutual collision and, hence, $T = \infty$. We can also derive from (3.2.5), with just one more *easy* step, an expression for the total distance S traveled by each bug, from the start of the cyclic pursuit until the mutual collision of all n bugs at C. We simply write $S = Tv$ which immediately gives us

(3.2.6)
$$S = \frac{r(0)}{\sin(\pi/n)}.$$

We derived (3.2.6) under the assumption that v is a constant but, in fact, the result is true even if $v = v(t)$, which I think is physically obvious after a little thought. We can formally show this if we derive S by integrating the differential arc length ds along a bug's path, that is, by writing

$$S = \int ds,$$

where, in polar coordinates, $(ds)^2 = (dr)^2 + (rd\theta)^2$:

$$ds = \sqrt{\left(\frac{dr}{d\theta}\right)^2 + r^2} d\theta.$$

From (3.2.4), which you'll recall is valid even if $v = v(t)$, we have

$$\frac{dr}{d\theta} = -r(0) \tan\left(\frac{\pi}{n}\right) \exp\left(-\theta \tan\left(\frac{\pi}{n}\right)\right),$$

and so

$$ds = \sqrt{r^2(0) \tan^2\left(\frac{\pi}{n}\right) \exp\left(-2\theta \tan\left(\frac{\pi}{n}\right)\right) + r^2(0) \exp\left(-2\theta \tan\left(\frac{\pi}{n}\right)\right)} d\theta$$

$$= r(0) \exp\left(-\theta \tan\left(\frac{\pi}{n}\right)\right) \sqrt{1 + \tan^2\left(\frac{\pi}{n}\right)} d\theta$$

$$= \frac{r(0) \exp\left(-\theta \tan(\pi/n)\right)}{\cos(\pi/n)} d\theta.$$

Since the cyclic pursuit begins with $\theta = 0$ and ends with $\theta = \infty$ (the bugs spiral around the origin an infinity of times before colliding at the center of the n-gon), we have

$$S = \frac{r(0)}{\cos(\pi/n)} \int_0^\infty \exp\left(-\theta \tan\left(\frac{\pi}{n}\right)\right) d\theta$$

$$= \frac{r(0)}{\cos(\pi/n)} \left\{ -\frac{\exp\left(-\theta \tan(\pi/n)\right)}{\tan(\pi/n)} \Bigg|_0^\infty \right.$$

$$= \frac{r(0)}{\cos(\pi/n)} \cdot \frac{1}{\tan(\pi/n)} = \frac{r(0)}{\sin(\pi/n)}.$$

This is (3.2.6), but here we have *not* made the constant speed assumption. Upon some reflection this is a plausible result, I think — if the bugs all obey the same $v(t)$ the total *distance* traveled shouldn't be affected (but, of course, the total time *will* be affected).

This remarkable result actually deserves a few more words of physical interpretation. When Martin Gardner presented his solution to the constant speed four-bug problem in *Scientific American*, he did so without using any high-powered (i.e., calculus) mathematics, arguing only (as did L. A. Graham earlier, in his 1959 book) that each bug is always on the side of a rotating, shrinking square (this follows by symmetry), and so "of course" each bug simply walks a total distance equal to the side length of the original square. For the Gardner/Graham problem, the square (a regular 4-gon) had a side length of one, which means $r(0) = \sqrt{2}/2$. Thus, (3.2.6) becomes, with $n = 4$,

$$S = \frac{\sqrt{2}/2}{\sin\left(\frac{\pi}{4}\right)} = \frac{\sqrt{2}/2}{1/\sqrt{2}} = \frac{2}{2} = 1,$$

and we see that the Gardner/Graham analysis does indeed give the correct answer. But this seductively simple and appealing argument *fails* in general, and is valid *only* in the $n = 4$ case! The reason it *is* true for $n = 4$ is because the angle between each of the four pairs of paths (each *pair* is the path of a pursuing bug and the path of the pursued bug) is always 90°, that is, the pursued bug has no speed component along the line-of-sight (LOS) to its pursuer, and so the closing speed between the two bugs is simply the speed of the pursuing bug. For any $n \neq 4$, however, a pursued bug *will* have a speed component along the LOS to its pursuer and this argument no longer applies.

As determined at the start of this section, the initial separation of any two adjacent bugs is $2r(0)\sin(\pi/n)$. This is the length of a side of the initial regular n-gon, and if we call this length $a(0)$, then

$$r(0) = \frac{a(0)}{2\sin(\pi/n)}$$

and so, from (3.2.6), we have the total distance traveled by each bug from start to finish as

(3.2.7) $$S = \frac{a(0)}{2\sin^2(\pi/n)}.$$

Evaluating (3.2.7) for some specific values of n, just to see how the numbers go:

n	S (in units of $a(0)$)
2	$\frac{1}{2}$
3	$\frac{2}{3}$
4	1
5	1.44721
6	2
7	2.65597
8	3.41421
100	506.7726

The $n = 2$ case applies to the case of two bugs at the ends of a line of length $a(0)$, and it is geometrically obvious that each will travel $a(0)/2$ to the center of the line; it is nice to see our formula in agreement with this.

There are several ways one can think to generalize the symmetrical n-bug problem. One way is to have each bug pursue not its nearest neighbor, but rather the bug that is k vertices away (in, say, the counterclockwise sense). The $k = 1$ case is the classic case we just analyzed — you can find the generalized results for integer $k > 1$ in Seery (1998). Another, more sophisticated generalization is to retain the nearest neighbor feature, but to have the n bugs chasing each other not on a plane but on a curved surface, such as a sphere. You can find the results for that twist in Aravind (1994).

Finally, here's a challenge problem for you to work on. As shown earlier, the total time, from start to finish, of the symmetric n-bug cyclic pursuit (on a plane) is, from (3.2.5),

$$T = \frac{r(0)}{v \sin(\pi/n)}.$$

When the chase is 99% over *in time*, how many times has each bug swung around the center of the n-gon? Evaluate your expression for the case of $n = 1,000$ bugs. Hint: integate (3.2.1) and (3.2.2) to get expressions for $r(t)$ and $\theta(t)$. You can find my calculations against which to check your answers in appendix G.

3.3 Morley's Nonsymmetrical 3-Bug Problem

This is the problem I mentioned in section 3.1, which appeared in Harry Bateman's 1918 textbook on differential equations. The problem is reduced in complexity in one sense (just three bugs, instead of the arbitrary n bugs), but also increased in complexity in another sense (the three bugs are initially on the vertices of an *arbitrary* triangle, not necessarily a regular 3-gon, an *equilateral* triangle). In the following analysis we will assume that all three bugs (A, B, and C) move with the same *constant* speed, which, with no loss of generality, we can take as unity with a suitable choice of units for time and/or distance. The geometry of the problem is illustrated in figure 3.3.1.

You'll recall that this problem is due to Frank Morley, who in particular was interested in the possibility of *non*-mutual capture in cyclic pursuit. In this section we'll see how to show that such a possibility is actually an *impossibility*, that is, for an arbitrary initial configuration, *three* bugs in a constant speed cyclic pursuit will *always* meet *simultaneously*. Furthermore, we'll also show that this meeting

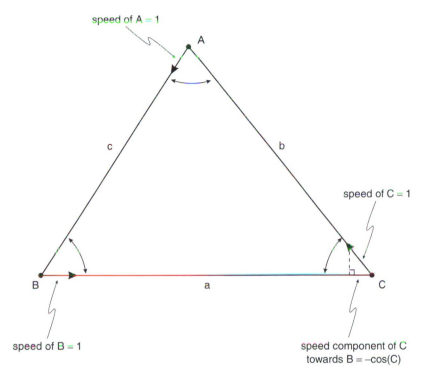

Figure 3.3.1 The geometry of Morley's non-symmetrical 3-bug problem

takes place after a finite time interval, a time interval on which we can place well-defined and fairly tight upper and lower bounds. The analysis that I'll show you is a mix of Bateman's approach and a modern discussion by Klamkin and Newman (1971), and it is, in my opinion, a stunning *tour de force*. Each individual step isn't particularly difficult, but there are a fair number of steps and each fits into the next one as tightly as a well-made tongue-and-groove floor joint.

To mathematically formulate Morley's problem, we start by writing the lengths of the sides of the triangle ABC as a, b, and c, where a is the length of the side opposite the similarly named bug (as shown in figure 3.3.1). I'll use A to also represent the *angle* of the vertex that locates bug A (and similarly for angles B and C). It should be clear from figure 3.3.1 that we can write the rates of change of the three side lengths as

$$(3.3.1) \qquad \frac{da}{dt} = -1 - \cos(C) = -[1 + \cos(C)],$$

$$(3.3.2) \qquad \frac{db}{dt} = -1 - \cos(A) = -[1 + \cos(A)],$$

and

$$(3.3.3) \qquad \frac{dc}{dt} = -1 - \cos(B) = -[1 + \cos(B)].$$

Since the cosine function is always in the interval -1 to $+1$, these three results immediately tell us that all three derivatives are *never* positive. You'll see in just a bit that this is an important observation — perhaps you'll say it's physically "obvious," since all three sides of the triangle formed by the three bugs should be continually getting shorter, but now we *know* that this is the case as well as just *how* the sides are shrinking. We'll use this observation to get to our final result.

From the law of cosines we have $a^2 = b^2 + c^2 - 2bc \cos(A)$ which, if we differentiate with respect to time (t), gives us

$$2a \frac{da}{dt} = 2b \frac{db}{dt} + 2c \frac{dc}{dt} - 2 \left(b \frac{dc}{dt} + c \frac{db}{dt} \right) \cos(A) + 2bc \sin(A) \frac{dA}{dt}$$

or, cancelling all the "2s,"

$$(3.3.4) \qquad a \frac{da}{dt} = b \frac{db}{dt} + c \frac{dc}{dt} - \left(b \frac{dc}{dt} + c \frac{db}{dt} \right) \cos(A) + bc \sin(A) \frac{dA}{dt}.$$

We can eliminate the derivatives of a, b, and c in (3.3.4) with the use of (3.3.1), (3.3.2), and (3.3.3), to arrive at

$$-a[1 + \cos(C)] = -b[1 + \cos(A)] - c[1 + \cos(B)]$$

$$+\{[b(1 + \cos(B)] + c[1 + \cos(A)]\} \cos(A)$$

$$+bc \sin(A) \frac{dA}{dt}$$

or, with a trivial rearrangement and expansion of terms,

$$bc\,\sin(A)\frac{dA}{dt} = -a[1+\cos(C)]+b[1+\cos(A)]$$

$$+c[1+\cos(B)]-b[1+\cos(B)]\cos(A)$$

(3.3.5)
$$-c[1+\cos(A)]\cos(A).$$

Now, applying the well-known[2] trigonometric identity

$$b\{\sin(A)\sin(B)-\cos(A)\cos(B)+1\} = c\{\cos(A)-\cos(B)\}$$

(3.3.6)
$$+a\{1+\cos(C)\}$$

to our arbitrary triangle, we can (perhaps surprisingly) greatly simplify (3.3.5). Using (3.3.6) to replace $a[1+\cos(C)]$ in (3.3.5), we have

$$bc\,\sin(A)\frac{dA}{dt} = -b\,\sin(A)\sin(B)+b\,\cos(A)\cos(B)-b+c\,\cos(A)$$

$$-c\,\cos(B)+b+b\,\cos(A)+c+c\,\cos(B)-b\,\cos(A)$$

$$-b\,\cos(B)\cos(A)-c\,\cos(A)-c\,\cos^2(A),$$

which becomes (after the amazing cancellation of all but *three* terms!)

$$bc\,\sin(A)\frac{dA}{dt} = -b\,\sin(A)\sin(B)+c-c\,\cos^2(A)$$

$$= -b\,\sin(A)\sin(B)+c-c\{1-\sin^2(A)\}$$

$$= -b\,\sin(A)\sin(B)+c\,\sin^2(A),$$

or, *at last,*

(3.3.7)
$$\frac{dA}{dt} = \frac{\sin(A)}{b} - \frac{\sin(B)}{c}.$$

Similar formulas apply for B and C, too, and they are easy to get from (3.3.7) by simply relabeling the vertices of our triangle. The labeling is, after all, completely arbitrary. For example, if we shift our labels one position clockwise, that is, $A \rightarrow C$, $B \rightarrow A$, $C \rightarrow B$, $a \rightarrow c$, $b \rightarrow a$, and $c \rightarrow b$, then (3.3.7) becomes

$$(3.3.8) \qquad \frac{dC}{dt} = \frac{\sin(C)}{a} - \frac{\sin(A)}{b},$$

and another clockwise shift results in

$$(3.3.9) \qquad \frac{dB}{dt} = \frac{\sin(B)}{c} - \frac{\sin(C)}{a}.$$

We are now in a position to show our first result — that a *non-simultaneous* capture of the bugs is impossible. We'll prove this by contradiction, that is, we'll show that the assumption of a nonsimultaneous capture leads to an obviously incorrect conclusion. What we'll specifically show is that the assumption of a nonmutual capture of B by A at time $t = t_0$ implies that at that instant $A + B + C < \pi$, while of course the fundamental geometry of the triangle formed by the three bugs demands that $A + B + C = \pi$ for *all t*. The false implication means the assumption of *non*-mutual capture must be incorrect. This is probably worth a few more words of elaboration. While we can argue that $A + B + C = \pi$ at every instant of time, it must also be true that *each angle individually* must be *less* than π, for example, $A = \pi$ and $B = C = 0$ is *not* a valid set of angles. This is so because these values clearly have the three bugs collinear, and it should be clear that that means the bugs have *always* been in a straight line (imagine time running backward — how could the bugs ever move *off* that straight line, no matter how far back in time we go?), which is in contradiction to our initial *triangular* configuration for the bugs. Thus,

$$(3.3.10) \qquad 0 < A, B, C < \pi.$$

Now, to start our demonstration of nonmutual capture, let's *assume* $c \rightarrow 0$ (A captures B) as $t \rightarrow t_{0-}$, while a and b do *not* $\rightarrow 0$ (B does *not* capture C at time t_0, and C does *not* capture A at time t_0), where "$t \rightarrow t_{0-}$" means t approaches t_0 from values *less* than t_0. Remember, t_0 is the supposed time at which A captures B, and *carefully* note that

$t_0 \neq 0$, that is, $t_0 > 0$. As argued earlier, (3.3.1), (3.3.2), and (3.3.3) together tell us that

$$(3.3.11) \qquad \frac{da}{dt} \leq 0, \quad \frac{db}{dt} \leq 0, \quad \frac{dc}{dt} \leq 0,$$

that is, a, b, and c each monotonically decrease and so, since all three quantities are physically bounded from below by zero (they are the sides of a triangle!), then all three must approach limiting values as $t \to \infty$. Thus, in addition to our assumed $\lim_{t \to t_{0-}} c = 0$, we must also have

$$\lim_{t \to t_{0-}} a = a_0 > 0$$

and

$$\lim_{t \to t_{0-}} b = b_0 > 0,$$

where the "greater than zero" conditions on a_0 and b_0 follow from our assumption that, as $t \to t_{0-}$, B does *not* capture C, and C does *not* capture A, respectively. (Note carefully that $a_0 \neq a(0)$, and $b_0 \neq b(0)$, i.e., $a(0)$ and $b(0)$ are the initial values of a and b.) From this it immediately follows that, for t sufficiently close to t_{0-}, we will have

$$(3.3.12) \qquad \frac{a}{b} - \frac{b}{c} < 0.$$

Now, from the law of sines, valid for any triangle, we have

$$\frac{a}{\sin(A)} = \frac{b}{\sin(B)},$$

or

$$\frac{a}{b} = \frac{\sin(A)}{\sin(B)},$$

and substituting this into (3.3.12) gives

$$\frac{\sin(A)}{\sin(B)} - \frac{b}{c} < 0,$$

or

(3.3.13)
$$\frac{\sin(A)}{b} < \frac{\sin(B)}{c}.$$

Looking back at (3.3.7), (3.3.13) tells us that, as $t \to t_{0-}$ (as our supposed capture of B by A approaches), we must see angle A decreasing. Since A is physically bounded from below by zero, we know that $\lim_{t \to t_{0-}} A$ exists and, whatever it may be, this limiting value is *less* than π by our earlier argument that A, B, and C must *all* be in the *open* interval 0 to π. So,

(3.3.14)
$$\lim_{t \to t_{0-}} A = A_0 < \pi.$$

Let's now turn our attention to what B and C do as $t \to t_{0-}$. First, B. For $0 < t < t_0$ it is certainly true that $a > a_0$ because each bug is always getting closer to its prey, and so from (3.3.9)

$$\frac{dB}{dt} > \frac{\sin(B)}{c} - \frac{\sin(C)}{a_0}$$

or, if we replace $\sin(C)$ with its greatest positive value in the interval for C — $(0,\pi)$ — we simply make the inequality stronger and arrive at

(3.3.15)
$$\frac{dB}{dt} > \frac{\sin(B)}{c} - \frac{1}{a_0}.$$

And since $\sin(B)/c > 0$ in the interval for B — $(0,\pi)$ — we make the inequality *even stronger* by dropping the first term on the right to get

$$\frac{dB}{dt} + \frac{1}{a_0} > 0,$$

which we can write as

(3.3.16)
$$\frac{d}{dt}\left(B + \frac{t}{a_0}\right) > 0.$$

The inequality of (3.3.16) says, in words, that $B + t/a_0$ is always increasing, and since it is bounded from above by $\pi + t_0/a_0$, then $\lim_{t \to t_{0-}} B$ must exist; let's call it B_0. Thus,

$$(3.3.17) \qquad\qquad \lim_{t \to t_{0-}} B = B_0.$$

Integrating (3.3.15) over the interval $0 \le t \le t_{0-}$, we have

$$\int_0^{t_{0-}} \frac{dB}{dt} dt = \int_0^{t_{0-}} dB > \int_0^{t_{0-}} \frac{\sin(B)}{c} dt - \int_0^{t_{0-}} \frac{1}{a_0} dt,$$

or

$$B(t_{0-}) - B(0) > \int_0^{t_{0-}} \frac{\sin(B)}{c} dt - \frac{t_{0-}}{a_0},$$

or, as $B(t_{0-}) = B_0$ from (3.3.17), we have

$$(3.3.18) \qquad\qquad B_0 - B(0) + \frac{t_{0-}}{a_0} > \int_0^{t_{0-}} \frac{\sin(B)}{c} dt.$$

The integral on the right-hand side of this inequality is clearly positive because the integrand is always positive over the entire interval of integration, and the left-hand side of (3.3.18) provides a well-defined upper bound to the integral. Thus, the integral surely exists, a conclusion that you'll see is *absolutely essential* to finishing our demonstration of simultaneous capture.

Now, recalling (3.3.3), we can change variables in the integral by $dt = -dc/(1 + \cos(B))$, and so the integral becomes

$$-\int_{c(0)}^{c_0} \frac{\sin(B)}{1 + \cos(B)} \cdot \frac{dc}{c},$$

or, since *by assumption* $c_0 = 0$, i.e., $\lim_{t \to t_{0-}} c = c_0 = 0$ because we are *assuming* A captures B at $t = t_0$, our integral becomes

$$\int_0^{c(0)} \frac{\sin(B)}{1 + \cos(B)} \cdot \frac{dc}{c}.$$

Then, recalling the identity $\sin(\theta)/(1 + \cos(\theta)) = \tan(\theta/2)$, we finally arrive at the following form for the integral:

$$\int_0^{c(0)} \frac{\tan(B/2)}{c} dc.$$

The penultimate step in our demonstration of mutual capture is, I think, beautiful in its elegant subtleness. We earlier showed that the above integral exists. The only way that can be is that, as $c \to 0$ in the denominator of the integrand (at the lower limit), we must also have $\tan(B/2) \to 0$ in the numerator of the integrand to keep the integral from diverging. If $\lim_{t \to t_{0-}} \tan(B/2) \neq 0$, then the integral would diverge logarithmically. Since we have also proven that $\lim_{t \to t_{0-}} B (= \lim_{c \to 0} B) = B_0$, this means that $\lim_{c \to 0} \tan(B/2) = \tan(B_0/2) = 0$. But *this* says

$$(3.3.19) \qquad \qquad \lim_{t \to t_{0-}} B = B_0 = 0.$$

To complete our analysis, we observe that *by assumption* c will eventually become the smallest side of the instantaneous bug triangle, and that means (by the law of sines) that C must become the smallest of the three angles. But since (3.3.19) says $B \to 0$ as $c \to 0$, then certainly so must $C \to 0$ as $c \to 0$, that is,

$$(3.3.20) \qquad \qquad \lim_{t \to t_{0-}} C = C_0 = 0.$$

And now we have our contradiction from (3.3.14), (3.3.19), and (3.3.20), which together say

$$\lim_{t \to t_{0-}} (A + B + C) < \pi,$$

which is clearly false as $A+B+C=\pi$, *always*. Thus, our initial assumption of a nonmutual capture must be false and we are done with our first result.

For our second result, that of establishing both upper and lower bounds on the capture time (t_c), we return to (3.3.1), (3.3.2), and (3.3.3), add them, and so arrive at

$$(3.3.21) \qquad \frac{da}{dt}+\frac{db}{dt}+\frac{dc}{dt}=-[3+\cos(A)+\cos(B)+\cos(C)].$$

Using the fact[3] that $A+B+C=\pi$, where all three angles are in the interval 0 to π, we can write

$$1 \leq \cos(A)+\cos(B)+\cos(C) \leq \frac{3}{2}.$$

Then (3.3.21) becomes the double inequality

$$-\frac{9}{2} \leq \frac{da}{dt}+\frac{db}{dt}+\frac{dc}{dt} \leq -4,$$

which, when integrated indefinitely, says

$$-\frac{9}{2}t \leq a(t)+b(t)+c(t)+k \leq -4t,$$

where k is the constant of indefinite integration. Setting $t=0$, we have

$$0 \leq a(0)+b(0)+c(0)+k \leq 0,$$

which says $a(0)+b(0)+c(0)+k=0$ (and so $k=-a(0)-b(0)-c(0)$), and thus

$$(3.3.22) \qquad -\frac{9}{2}t \leq a+b+c-a(0)-b(0)-c(0) \leq -4t.$$

At the instant of mutual capture, at $t = t_c$, we have (by definition of *capture*) $a = b = c = 0$. So,

$$\frac{9}{2}t_c \geq a(0) + b(0) + c(0) \geq 4t_c.$$

Therefore, at last,

(3.3.23)
$$\frac{a(0) + b(0) + c(0)}{9/2} \leq t_c \leq \frac{a(0) + b(0) + c(0)}{4}.$$

We can partially check the validity of this result by looking back at (3.2.7), our expression for the total distance S traveled by each of n bugs starting initially on the vertices of a regular n-gon. For $n = 3$ bugs starting on the vertices of an equilateral triangle (so $a(0) = b(0) = c(0)$), we have

$$S = \frac{a(0)}{2\sin^2(\pi/3)} = \frac{a(0)}{3/2} = \frac{2}{3}a(0).$$

For the *unit* speed bugs we have in our present analysis, then, $t_c = \frac{2}{3}a(0)$. On the other hand, our double inequality of (3.3.23) says

$$\frac{3a(0)}{9/2} \leq t_c \leq \frac{3a(0)}{4},$$

or

$$\frac{2}{3}a(0) \leq t_c \leq \frac{3a(0)}{4},$$

an inequality that, as we now see, includes the *exact* value of t_c as its lower bound.

With the 1971 proof by Klamkin and Newman that three equal speed bugs, no matter their initial positions, *always* capture one another simultaneously, it is natural to ask what can be said for $n > 3$ bugs. Klamkin and Newman speculated that the "always mutual capture" feature of the $n = 3$ case could be extended to all n, but that proved not to be so. Behroozi and Gagnon (1979) — the sequel to their 1975 computer-assisted study of cyclic pursuit — showed that $n = 4$

equal speed bugs will indeed always have mutual, simultaneous capture *if* the initial bug positions are the vertices of a *convex* quadrilateral. In this same paper, however, they showed a specific example of *non*-simultaneous capture for $n = 4$ equal speed bugs starting from the vertices of a *non*-convex quadrilateral. And finally, Richardson (2001) extended the possibility of *non*-simultaneous capture to *all* $n > 3$, with numerous specific computer-simulated examples for $n = 5$, 6, and 8.

Chapter 4

Seven Classic
Evasion Problems

4.1 The Lady-in-the-Lake Problem

This is a clever pursuit-and-evasion problem with the emphasis on *evasion*. At its most elementary level it became famous decades ago when, like the four-bug cyclic pursuit problem in the previous chapter, it appeared in Martin Gardner's "Mathematical Games" column in *Scientific American* (November and December 1965). As Gardner presented the problem then to his readers,

> A young lady was vacationing on Circle Lake, a large artificial body of water named for its precisely circular shape. To escape from a man who was pursuing her, she got into a rowboat and rowed to the center of the lake, where a raft was anchored. The man decided to wait it out on shore. He knew she would have to come ashore eventually; since he could run four times faster than she could row, he assumed that it would be a simple matter to catch her as soon as her boat touched the lake's edge. But the girl — a mathematics major at Radcliffe — gave some thought to her predicament. She knew that on foot she could outrun the man

[which does raise the question of why such a smart lady got herself into this situation in the first place by rowing out into a lake!]; it was only necessary to devise a rowing strategy that would get her to a point on shore before *he* could get there. She soon hit on a simple plan, and her applied mathematics applied successfully. What was the girl's strategy?

This problem can be generalized into one with just a bit more sophistication, and I'll do that before we are through, but for now let's concentrate on Gardner's specific case. For the particular speed ratio of four there is, *perhaps* surprisingly, a solution that requires almost no mathematics at all.

The lady's escape strategy consists of two stages. She first hops into her boat and rows away from the raft in such a way that she, the raft, and the man are always collinear. This first part of the lady's rowing path will clearly have to change direction constantly to continually maintain collinearity because the man will instantly begin running around the lake's edge in his attempt to intercept her at the shore. This is illustrated in figure 4.1.1, where I've assumed the man runs *counterclockwise* around the lake (he could, of course, just as well have chosen to run clockwise — the choice of his running direction is completely arbitrary). As I'll next show you, the lady can indeed maintain collinearity — or at least she can *for a while*.

To see why there is a limit on the first stage of the lady's escape strategy (whereupon she'll shift, perhaps it goes without saying but I'll say it anyway, to her *second* stage strategy), let's assume that the man runs at speed v and that the lady rows with speed αv. Thus, in the original statement of the problem $\alpha = 0.25$. We see from figure 4.1.1 that the man opens up the angle θ at the rate of $d\theta/dt = v/R$, where θ is measured with respect to the line initially joining the man and the lady-on-the-raft. Without any loss of generality we can assume that this initial line is the vertical axis of our coordinate system, as shown in figure 4.1.1. Since the lady's *angular* speed component must be

$$v_\theta = r \frac{d\theta}{dt}$$

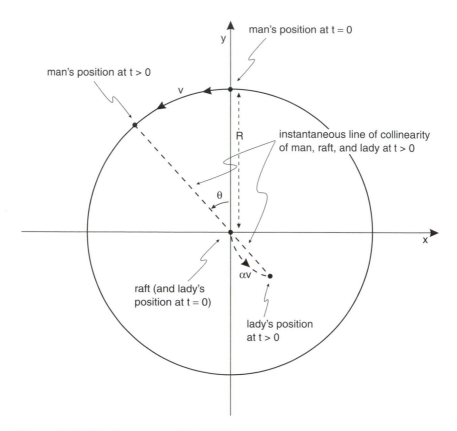

Figure 4.1.1 The first stage of the lady's escape

for her to maintain the raft between herself and the man, we can write her angular speed as

$$(4.1.1) \qquad\qquad v_\theta = v\frac{r}{R}.$$

The farther she gets from the raft, then, (4.1.1) tells us the greater must be her angular speed if she is to maintain collinearity.

Now, since the lady's *total* speed through the water is αv, her *radial* speed component (v_r) must be such that

$$v_r^2 + v_\theta^2 = (\alpha v)^2,$$

because her total speed is geometrically represented by the hypotenuse of a right triangle, with perpendicular sides v_r and v_θ. Thus,

$$v_r = \sqrt{\alpha^2 v^2 - v_\theta^2} = \sqrt{\alpha^2 v^2 - v^2 \frac{r^2}{R^2}},$$

or,

(4.1.2)
$$v_r = \frac{dr}{dt} = v\sqrt{\alpha^2 - \frac{r^2}{R^2}}.$$

The lady has a positive v_r (that is, she moves ever closer to shore, all the while keeping half the lake between herself and the man) as long as $\alpha^2 - r^2/R^2 > 0$, that is, until $r = \alpha R$. At the instant her v_r drops to zero she switches to the second stage of her escape strategy, which I'll describe in just a moment.

But first, *how long* does it take her to arrive at the condition $v_r = 0$? Since $dt = dr/v_r$, then if we call $t = T$ the time at which $v_r = 0$, we have

$$\int_0^T dt = T = \int_0^{\alpha R} \frac{dr}{v_r} = \int_0^{\alpha R} \frac{dr}{v\sqrt{\alpha^2 - r^2/R^2}} = \frac{R}{v} \int_0^{\alpha R} \frac{dr}{\sqrt{(\alpha R)^2 - r^2}}$$

$$= \frac{R}{v} \left\{ \sin^{-1}\left(\frac{r}{\alpha R}\right) \Big|_0^{\alpha R} = \frac{R}{v} \sin^{-1}(1), \right.$$

or

(4.1.3)
$$T = \frac{\pi R}{2v}.$$

When the lady arrives at the circle with radius αR centered on the raft, at time $t = T$, she has arrived at what I call the "go-for-broke" circle. I call it that because then, now that she is no longer moving ever closer to shore with the first part of her escape strategy, she forgets about maintaining collinearity and simply rows straight for shore at her full water speed of αv. She has distance $R - \alpha R$ to row (at speed αv) and the man has distance πR (half the circumference of the lake) to run at speed v. She gets to shore before he gets to her *if*

$$\frac{R - \alpha R}{\alpha v} < \frac{\pi R}{v},$$

or

$$R(1 - \alpha) < \pi\alpha R,$$

or

$$1 - \alpha < \pi\alpha,$$

or

$$1 < \alpha(1 + \pi),$$

or, at last, *if*

(4.1.4) $$\alpha > \frac{1}{1 + \pi} = 0.241453.$$

Since $\alpha = 0.25$ in the *Scientific American* version of the problem, we see that this two-stage escape strategy (just barely) works and that the lady's virtue is (just barely) preserved.

Of course, if α is sufficiently large there is no need for a *two*-stage escape strategy. It is easy to see that if α is "big enough" then all the lady needs to do is immediately row directly to shore, to the point directly opposite the man's location. She gets to shore before he gets to her *if*

$$\frac{R}{\alpha v} < \frac{\pi R}{v},$$

that is, if $\alpha > 1/\pi = 0.3183099\ldots$. Still, while not essential for her escape, the two-stage strategy will give the lady a little extra head start on the man, and it is interesting to calculate how much this head start is for $\alpha = 1/\pi$. As before, in the two-stage strategy the man, the raft, and the lady remain collinear until the lady reaches the go-for-broke circle, with radius $\alpha R = R/\pi$. *Then* she rows straight for shore, now distance $R - R/\pi = R(1 - 1/\pi)$ away. Since her rowing speed is $\alpha v = v/\pi$, this requires a time (during her *second* stage) of

$$\frac{R(1 - 1/\pi)}{v/\pi} = \frac{R}{v}(\pi - 1).$$

The man reaches her landing point on the shore after running halfway around the lake, which requires a time (starting at the instant the lady "goes for broke") of

$$\frac{\pi R}{v} = \frac{R}{v}\pi.$$

So, she arrives at her landing point on shore *before* he does by a time interval of

$$\frac{R}{v}\pi - \frac{R}{v}(\pi - 1) = \frac{R}{v}.$$

To put this head start (in time) in perspective, it is the time it takes the man to run distance R, the radius of the lake; a not insignificant head start.

Let's suppose now that the lady does *not* have a "big" α. Suppose, in fact, that it is smaller than $(1+\pi)^{-1}$. Is it then impossible for her to escape from the man? Actually, if we make a plausible assumption about the man's reasoning (specifically, that he is *rational*), then it is *still possible* for a slow-rowing lady to escape. Since the lady is a Radcliffe math major, it seems likely she originally met the ungentlemanly lout while attending a math course, which means *he* surely knows some mathematics, too (even louts can learn to add). Therefore, let's assume that, as soon as the lady leaves the raft and begins to execute the first stage of her escape strategy, the man deduces what she is up to. That is, he observes that as *he* moves, *she* moves to keep the raft between him and her even as she moves ever closer to shore. He then further deduces that as soon as she reaches her go-for-broke circle she will head straight for shore. So, here's our assumption — as soon as he sees her go to the second stage of her escape strategy, that is, at the instant she makes straight for shore, he stops watching her carefully and simply runs around the lake to the point on shore where he now "knows" she is heading. The only thing that will cause him to reevaluate matters is if the lady stops her go-for-broke rowing and, for whatever reason, begins to move back toward the raft.

Being a clever Radcliffe math major, however, and knowing her α is less than $(1+\pi)^{-1}$, she has one last trick up her sleeve. She will,

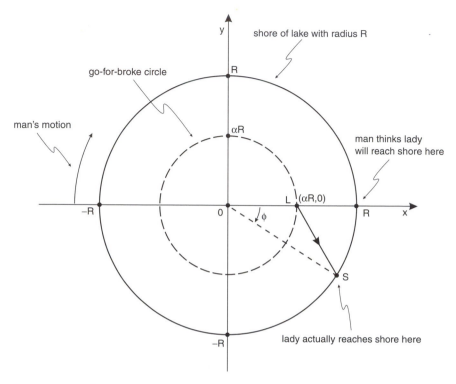

Figure 4.1.2 The way things are when the lady reaches her "go-for-broke" circle

indeed, row a straight-line path to shore as soon as she reaches her go-for-broke circle, but it will *not* be the shortest-distance straight-line path that the man thinks she will row. To see what she has in mind instead, look at figure 4.1.2, where, with no loss in generality, I've put the lady's position at the instant she reaches her go-for-broke circle at $(\alpha R, 0)$. The man's position at that instant is, of course, $(-R, 0)$. In the notation of the figure, ϕ is the angle the straight line joining the raft to the lady's landing point on shore (S) makes with the horizontal axis. The man is assuming that $\phi = 0$, but he is wrong, as you'll soon see.

Let's slowly work our way up to understanding the lady's new escape strategy. First, we will greatly simplify our calculations by noticing that the ratio of the radius of the go-for-broke circle to the radius of the lake is simply $\alpha R/R = \alpha$. If we next denote the radius of the go-for-broke circle as our *unit* distance (which we can do with no loss of generality), then $\alpha R = 1$, and so

(4.1.5)
$$\alpha = \frac{1}{R}.$$

What this means is that if we wish to find the *smallest* value for α for which the lady can still escape, then an equivalent problem is that of finding the *largest* R for which the lady can still escape. And finally, since the lady rows at speed αv, we can write her rowing speed as $(1/R)v = v/R$. We can now set the problem up mathematically as follows.

When the lady reaches her go-for-broke circle (point L in the figure) she is distance $\alpha R = 1$ from the raft, and the law of cosines tells us that the distance she has left to row to shore to reach point S is

$$LS = \sqrt{1 + R^2 - 2R\cos(\phi)}.$$

This takes her a time interval of

(4.1.6)
$$\frac{LS}{v/R} = \frac{\sqrt{1 + R^2 - 2R\cos(\phi)}}{v/R} = \frac{R}{v}\sqrt{1 + R^2 - 2R\cos(\phi)}$$

to row.

What will the man do? As I've drawn figure 4.1.2, the man is running clockwise around the lake to S — but, you may object, with ϕ as I've drawn it wouldn't it make more sense for him to run *counterclockwise*, because then his running distance to S is shorter? Here's where we have to again make the assumption of rationality that I mentioned earlier (even louts can be rational). I'll simply quote Schuurman and Loder (1974) about what both the lady and the man conclude once she reaches her go-for-broke circle:

> ... she performs an infinitesimal radial feint [toward the shore that leads the man to start running clockwise]. From that moment on, [the man's] best policy is to continue running clockwise if [the lady] goes to shore along a straight line (always better than a curved one!) *not crossing the* [go-for-broke] *circle*. If [the man] would return, a new diametrical mutual position, advantageous to [the lady] would be established.

This last sentence is important to understand — what it points out is that if the man should at any time reverse his running direction around the lake *then* the lady could, *at the least*, start rowing directly away from him at the instant of his reversal and head straight for shore. That would have her starting the second stage of her original escape strategy from a point *beyond* the go-for-broke circle, and yet still leave the man with half the lake's circumference to travel. Even better (from the lady's point of view), of course, would be for her to simply flip the sign of ϕ, and then the situation is just as it was before *he* switched! So, once the man has committed to a running direction we see that he gains nothing by reversing his decision — he therefore (if rational) will run through the angle $\pi + \phi$ to reach S. I'll elaborate on the reason for the lady to choose a straight line rowing path *not crossing* the go-for-broke circle in just a moment.

The time required for the man to run a distance $(\pi + \phi)R$ around the lake to S is, of course,

$$(4.1.7) \qquad \frac{R}{v}(\pi + \phi).$$

Thus, the lady will just escape his clutches if the two times given by (4.1.6) and (4.1.7) are equal, that is, if

$$\pi + \phi = \sqrt{1 + R^2 - 2R\cos(\phi)}.$$

Squaring both sides and solving for R gives

$$R = \cos(\phi) \pm \sqrt{\cos^2(\phi) + (\pi + \phi)^2 - 1},$$

and since $R > 0$ (it is the *physical* radius of a lake) we must use the plus sign,

$$(4.1.8) \qquad R = \cos(\phi) + \sqrt{\cos^2(\phi) + (\pi + \phi)^2 - 1}.$$

Figure 4.1.3 shows the behavior of $R(\phi)$, and it is obviously a *non-decreasing* function of ϕ. To find the *smallest* α for which the lady escapes we must use the *largest* possible value for R — look back at (4.1.5). That is, we want to find the value of ϕ that maximizes $R(\phi)$.

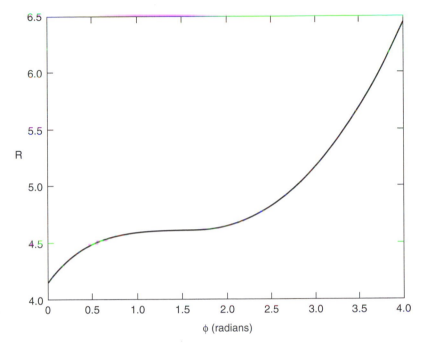

Figure 4.1.3 The radius of the lake

Now, even though R continually gets bigger (certainly it never decreases) with increasing ϕ, there *is* a limit on how big ϕ can be. If ϕ exceeds the value it has such that the line LS (in figure 4.1.2) is *tangent* to the go-for-broke circle, then the lady's rowing path will take her back *inside* the go-for-broke circle, that is, she will have a radial speed component pointing *back toward the raft* (which is not a feature we expect to see in an *escape* strategy). That is, the lady should pick ϕ such that the line LS is perpendicular to the x-axis. From figure 4.1.2 we see that this value of ϕ $(= \phi_t)$ satisfies the condition (don't forget, we have taken αR as unity)

$$\cos(\phi_t) = \frac{\alpha R}{R} = \frac{1}{R}.$$

If we substitute this condition in (4.1.8) we get

$$\frac{1}{\cos(\phi_t)} = \cos(\phi_t) + \sqrt{\cos^2(\phi_t) + (\pi + \phi_t)^2 - 1},$$

which, with just a few easy steps of algebra, reduces to the transcendental equation

(4.1.9) $\tan(\phi_t) = \pi + \phi_t.$

It is clear, simply by sketching the curves for each side of (4.1.9), that there is a solution to (4.1.9) somewhere in the interval $(0, \pi/2)$. We can solve for this value by numerical means,[1] with the result $\phi_t = 1.3518168\ldots$ radians, which gives

$$\cos(\phi_t) = \frac{1}{R_{\max}} = \alpha_{\min} = 0.2172336\ldots.$$

Alternatively, the lady can escape even if the man runs $1/\alpha_{\min} = 4.6033388\ldots$ times as fast as she can row, which is significantly greater than the factor of four given in the *Scientific American* version of the problem.

4.2 Isaacs's Guarding-the-Target Problem

There is a very pretty, *general* pursuit-and-evasion problem in the classic 1965 book (reprinted by Dover in 1999) *Differential Games* by the American mathematician Rufus Isaacs (1914–1981). A related problem, using game theoretic ideas (discussed later in this chapter), is in Ruckle (1979). Isaacs states his problem, along with giving its *general* solution, as follows (with reference to figure 4.2.1):

> Both **P** and **E** [pursuer and evader] travel with the same speed.... The motive of **P** is to guard a target **C**, which we take as an area in the plane, from attack by **E**.... The optimal strategies [for both **P** and **E**] are found thus: Draw the perpendicular bisector of **PE** [where **P** and **E** denote starting positions]. Any point in the half-plane above this line can be reached by **E** prior to **P**, and this property fails in the lower half-plane. Clearly **E** should head for the best of his accessible points. Let **D** be the point of the bisector nearest **C**. The optimal strategies for both [**P** and **E**] decree that they travel toward **D**. Capture occurs there...

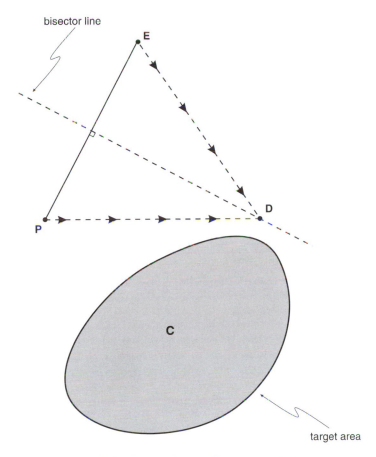

bisector line

E

D

P

C

target area

Figure 4.2.1 **P** and **E** defending and attacking, respectively, the target area **C**

In Isaacs's miltary conception of this problem, **E** must actually reach *at least* the boundary of **C** to be successful in its attack. For **E** to penetrate to an interior point of **C** is even "better," of course, but not necessary. **P** is successful in defeating **E** if the capture point **D** is anywhere outside **C**. It is difficult to say much more in general than this about the problem, *unless* we are given some additional information about the specifics of **C**, that is, its shape and dimensions. If we simultaneously *simplify* the structure of **C** and introduce a more specific definition of *capture*, however, then we *can* derive some specific results. For example, let's suppose we take the viewpoint of the evader **E** that is attempting to destroy the target that is being guarded by **P**, and that **C** is a *point* target, for example, the location of a specific enemy commander or of an enemy radio transmitter. Furthermore, let's suppose that **E** is carrying an explosive device which, when detonated, has a circular radius of destruction of $R > 0$. And finally, let's suppose that **P** can stop **E** only by direct impact, that is, **P** must actually intercept **E**. Then, for **E** to be successful, all it must do is come within a distance less than or equal to R *before* **P** reaches **E**. This might, for example, be a crude first-order model for an attacking missile versus a missile defense system that is supposed to protect an area (e.g., a city) against a ballistic missile attack. With no loss in generality we can take **P** at time $t = 0$ at the origin of an x–y coordinate system, and the *point* target **C** on the x-axis at $x = x_c > 0$. That is, the target is to the right of **P**. The case for a target initially to the left of **P** is simply the mirror-image of our assumed case, and so offers no new complications. In this system let **E** be at (x_0, y_0) at time $t = 0$. All of this is illustrated in figure 4.2.2.

Note, carefully, that if **C** is one of **E**'s accessible points then the problem is trivial, because **E** can *always* destroy **C** by actually *reaching* **C** before **P** can reach **E**. So, we'll further assume that **C** is not one of **E**'s accessible points, an assumption equivalent to assuming that y_0 is "sufficiently large."

The equation of the bisector line L_1 is easily shown to be

$$(4.2.1) \qquad y = -\frac{x_0}{y_0}x + \frac{x_0^2 + y_0^2}{2y_0},$$

which has the required slope and does, indeed, pass through the point midway between **P** and **E** at time $t = 0$, that is, through the point $(x_0/2, y_0/2)$. Now, L_2 is the *perpendicular* line segment from the point target **C** to L_1, and so the length of L_2 is the closest approach distance of **E** to **C**. The slope of L_2 is clearly y_0/x_0 and, since it passes through the point $(x_c, 0)$, it is easy to see that the equation of L_2 is

$$(4.2.2) \qquad y = \frac{y_0}{x_0}x - \frac{y_0}{x_0}x_c.$$

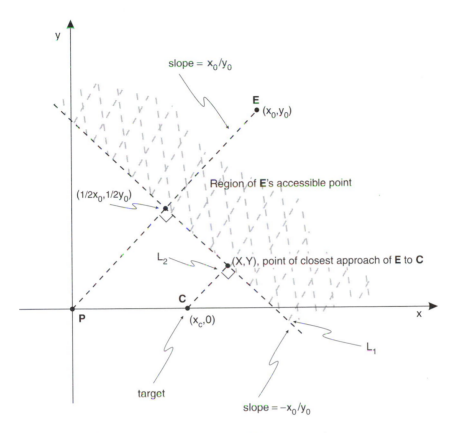

Figure 4.2.2 The geometry of Isaacs's problem for a point target

The point of closest approach of **E** to **C** is the intersection of L_1 and L_2, the point (X,Y). From (4.2.1) and (4.2.2) we have

$$\frac{y_0}{x_0}X - \frac{y_0}{x_0}x_c = -\frac{x_0}{y_0}X + \frac{x_0^2 + y_0^2}{2y_0},$$

which is quickly solved to give

$$(4.2.3) \qquad X = \frac{y_0^2}{x_0^2 + y_0^2}x_c + \frac{x_0}{2}.$$

Inserting (4.2.3) into either (4.2.1) or (4.2.2) then gives

$$(4.2.4) \qquad Y = \frac{y_0}{x_0}\left(\frac{y_0^2}{x_0^2 + y_0^2}x_c + \frac{x_0}{2}\right) - \frac{y_0}{x_0}x_c.$$

The length (squared) of L_2 is $(X - x_c)^2 + Y^2$, or

$$\left[\frac{y_0^2}{x_0^2 + y_0^2}x_c + \frac{x_0}{2} - x_c\right]^2 + \left[\frac{y_0}{x_0}\left(\frac{y_0^2}{x_0^2 + y_0^2}x_c + \frac{x_0}{2}\right) - \frac{y_0}{x_0}x_c\right]^2,$$

which, after a bit of algebra that I encourage you to verify, reduces to

$$\frac{[x_0(x_0^2 + y_0^2) - 2x_c x_0^2]^2}{4x_0^2(x_0^2 + y_0^2)}.$$

For **E** to achieve its mission goal of destroying **C**, the circular radius of destruction R (squared) of **E**'s weapon must exceed this, that is,

$$R^2 > \frac{[x_0(x_0^2 + y_0^2) - 2x_c x_0^2]^2}{4x_0^2(x_0^2 + y_0^2)},$$

or

$$R > \frac{x_0(x_0^2 + y_0^2) - 2x_c x_0^2}{2x_0\sqrt{x_0^2 + y_0^2}}.$$

We can normalize this result with respect to x_c as follows.

$$R > x_c \frac{x_0/x_c(x_0^2 + y_0^2) - 2x_0^2}{2x_0\sqrt{x_0^2 + y_0^2}}$$

and so

$$\frac{R}{x_c} > \frac{(x_0^2 + y_0^2)/x_c - 2x_0}{2\sqrt{x_0^2 + y_0^2}} = \frac{(x_0^2 + y_0^2)/x_c - 2x_0}{2x_c\sqrt{(x_0^2 + y_0^2)/x_c^2}},$$

or, at last,

$$(4.2.5) \qquad \frac{R}{x_c} > \frac{(x_0/x_c)^2 + (y_0/x_c)^2 - 2(x_0/x_c)}{2\sqrt{(x_0/x_c)^2 + (y_0/x_c)^2}}.$$

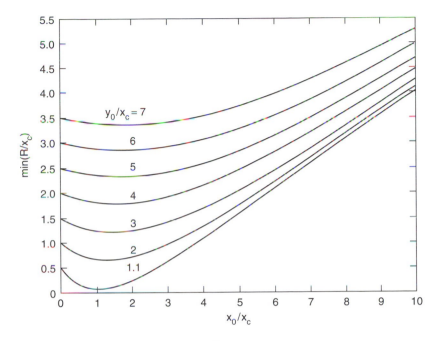

Figure 4.2.3 The minimum value of $\frac{R}{x_c}$ for a given value of $\frac{y_0}{x_c}$

Figure 4.2.3 shows several of the curves that represent the right-hand side of (4.2.5); each curve is the minimum R/x_c, as a function of x_0/x_c, for a given fixed value of y_0/x_c (the label-value next to each curve). From these curves **E** can determine the minimum value of R (the *amount* of explosive) required for success in destroying **C** as a function of both **E**'s starting point and the location of the target.

4.3 The Hiding Path Problem

A fascinating evasion problem is discussed in Bailey (1994), involving a clever rabbit who is trying to avoid detection by a nearby fox. The geometry of the problem is shown in figure 4.3.1, which has the fox initially at $(0, -b)$ and running to the left at constant speed V_f along a straight-line path, while the rabbit is initially at $(0, c)$ and running to the right at its constant, full-steam-ahead speed of V_r. The values of b

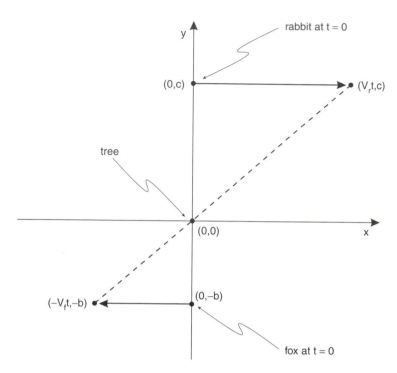

Figure 4.3.1 The geometry of the hiding path problem for the special case where the rabbit runs along a straight, horizontal path

and c are both non-negative, just as shown in the figure. There is a tree at $(0,0)$ and so, *initially at least*, the rabbit is not in the line-of-sight of the fox. The rabbit's problem is that of determining the path (called the *hiding path*) on which to run — not necessarily a straight-line path, as you'll soon see — so as to *always* keep the tree between itself and the fox.

Before showing you Bailey's solution, let me make some brief historical remarks. First, I found it nicely appropriate to have Professor Bailey, in particular, write on this problem because he was (and still is, as I write) Professor Emeritus of Mathematics at Rose-Hulman Institute of Technology in Terre Haute, Indiana, which is the renamed Rose Polytechnic Institute, where, you'll recall, Professor A. S. Hathaway (of dog vs. duck fame in chapter 2) once hung his hat. Professor Bailey followed in historically important footsteps. Still, while I will likewise follow in Bailey's footsteps here, the hiding path problem did not originate with him. It can be found, decades earlier, in Wilder (1931), posed as follows:

> An automobile moves along a straight road with a constant speed (v) while a man in a field beside the road walks with constant speed (u) along such a path as to keep a tree between him and the automobile. Determine his path.

But the problem is not Wilder's, either — as he stated in the opening paragraph of his essay, the problem was "proposed to Professor B. H. Brown by some doubtless well-meaning freshman"! (Wilder and Brown were both professors of mathematics at Dartmouth College, New Hampshire, in the first half of the twentieth century. The identity of the "well-meaning freshman," alas, has been lost to history.) All right, back to Professor Bailey's solution.

In figure 4.3.1 I've drawn one very special solution to the rabbit's problem. In that figure I've assumed that the rabbit's hiding path is a straight-line path — remember, the *fox's* path is *given* as a straight, horizontal path. From similar triangles we have, for all $t \geq 0$,

$$\frac{V_r t}{c} = \frac{V_f t}{b}$$

or, $V_r = c V_f / b$. This is a very precise requirement, of course; what if V_r is either less than or greater than $c V_f / b$? The answer is that the rabbit may still be able to hide, but its path in general will be not straight but curved.

Figure 4.3.2 shows the general case of the rabbit running at constant speed V_r along a curved path, where I'll write $s(t)$ as the distance the rabbit has run at time t ($s(0) = 0$). The differential arc length ds along the rabbit's hiding path satisfies $(ds)^2 = (dx)^2 + (dy)^2$ and so

$$\left(\frac{ds}{dt}\right)^2 = \left(\frac{dx}{dt}\right)^2 + \left(\frac{dy}{dt}\right)^2,$$

or, as $V_r = ds/dt$,

$$(4.3.1) \qquad V_r^2 = \left(\frac{dx}{dt}\right)^2 + \left(\frac{dy}{dt}\right)^2.$$

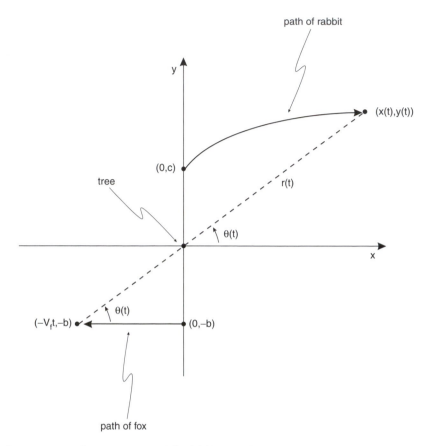

Figure 4.3.2 The geometry of the hiding path problem in the general case where the rabbit runs along a curved path

Using the geometry of similar triangles once more, figure 4.3.2 also tells us that

$$\tan\{\theta(t)\} = \frac{y(t)}{x(t)} = \frac{b}{V_f t},$$

or

(4.3.2) $$x(t) = \frac{V_f t}{b} y(t).$$

Differentiation of (4.3.2) gives

(4.3.3) $$\frac{dx}{dt} = \frac{V_f}{b}\left[t\frac{dy}{dt} + y\right],$$

which, when substituted into (4.3.1), gives

$$V_r^2 = \left(\frac{V_f}{b}\right)^2 \left[t\frac{dy}{dt} + y\right]^2 + \left(\frac{dy}{dt}\right)^2.$$

After just a bit of easy algebra, this becomes

$$(4.3.4) \quad V_r^2 = \left[1 + \left(\frac{V_f t}{b}\right)^2\right]\left(\frac{dy}{dt}\right)^2 + 2\left(\frac{V_f}{b}\right)^2 ty\frac{dy}{dt} + \left(\frac{V_f}{b}\right)^2 y^2.$$

To keep from being overwhelmed by the notation, let's make the following *normalization* definitions:

$$(4.3.5) \qquad X = \frac{V_f}{bV_r}x, \quad Y = \frac{V_f}{bV_r}y, \quad T = \frac{V_f t}{b},$$

where you'll surely notice that X, Y, and T are all *dimensionless*. These definitions are useful because we can then write, with the help of the chain rule from calculus,

$$\frac{dy}{dt} = \frac{d}{dt}\left(\frac{bV_r Y}{V_f}\right) = \frac{d}{dT}\left(\frac{bV_r Y}{V_f}\right)\cdot\frac{dT}{dt} = \frac{bV_r}{V_f}\cdot\frac{dY}{dT}\cdot\frac{V_f}{b},$$

or

$$(4.3.6) \qquad\qquad \frac{dy}{dt} = V_r\frac{dY}{dT}.$$

Substituting (4.3.5) and (4.3.6) into (4.3.4), we have

$$V_r^2 = (1 + T^2)V_r^2\left(\frac{dY}{dT}\right)^2 + 2\left(\frac{V_f}{bV_r}yV_r\right)\left(\frac{V_f t}{b}\right)\left(V_r\frac{dY}{dT}\right) + \frac{V_f^2}{b^2}\cdot\frac{b^2 V_r^2}{V_f^2}Y^2,$$

or

$$1 = (1 + T^2)\left(\frac{dY}{dT}\right)^2 + 2YT\frac{dY}{dT} + Y^2,$$

or

$$(4.3.7) \qquad (1 + T^2)\left(\frac{dY}{dT}\right)^2 + 2YT\frac{dY}{dT} + Y^2 - 1 = 0.$$

This may look pretty awful — and I must admit it isn't the prettiest equation you'll ever see! — but once you observe that (4.3.7) is a quadratic in dY/dT it becomes much less ugly. Solving for dY/dT gives

$$\frac{dY}{dT} = \frac{-2YT \pm \sqrt{4Y^2T^2 - 4(1+T^2)(Y^2-1)}}{2(1+T^2)}$$

or

(4.3.8)
$$\boxed{\frac{dY}{dT} = \frac{-YT \pm \sqrt{1+T^2-Y^2}}{1+T^2}},$$

a result which I've put in a box to emphasize its importance — we'll be using this differential equation for Y *a lot*.

We can derive a similar differential equation for X as follows. Recalling (4.3.2) and our definition for X, we see that

$$X = \frac{V_f}{bV_r}x = \frac{V_f}{bV_r} \cdot \frac{V_f t}{b}y = \frac{V_f}{bV_r}y \cdot \frac{V_f t}{b}$$

which reduces to simply

(4.3.9)
$$X = YT.$$

Differentiating (4.3.9) with respect to T, and then using (4.3.8), gives

$$\frac{dX}{dT} = Y + T\frac{dY}{dT} = Y + \frac{-YT^2 \pm T\sqrt{1+T^2-Y^2}}{1+T^2},$$

or

(4.3.10)
$$\frac{dX}{dT} = \frac{Y \pm T\sqrt{1+T^2-Y^2}}{1+T^2}.$$

Our next logical step would be to attempt to solve the two coupled differential equations (4.3.8) and (4.3.10) for X and Y, but it isn't at all clear that closed-form expressions for X and Y exist. At the end of this section we'll settle for numerical, computer-generated solutions for possible hiding paths, but there is a *lot* of information in (4.3.8) and (4.3.10) that we can discover *without* solving those equations. To begin, notice that since the constants b and c (whose values determine the initial locations of the fox and the rabbit, respectively) are both non-negative, then the initial value of Y is also non-negative, that is,

$$Y(0) = \frac{V_f}{bV_r}y(0) = \frac{V_f}{bV_r}c \geq 0.$$

Now, take a look at (4.3.8) and (4.3.10) and notice that both equations involve the radical $\sqrt{1+T^2-Y^2}$. This means that at time $t = 0$ ($T = 0$) we must have $Y \leq 1$ to keep the radical from being imaginary because, after all, both dX/dT and dY/dT have immediate physical significance — normalized *speeds* — that requires them to be *real* quantities. So, without solving any equations, we have determined that Y obeys the constraint

(4.3.11) $$0 \leq Y(0) \leq 1.$$

This, in turn, tells us that

$$Y(0) = \frac{V_f}{bV_r}c \leq 1,$$

or

$$V_r \geq \frac{V_f}{b}c.$$

In other words, if the rabbit isn't *at least* as fast as $V_f c/b$, then it is not possible for the rabbit to hide from the fox.

Another interesting issue is related to the choice of algebraic sign (plus or minus) in front of the radicals in (4.3.8) and (4.3.10). If we choose the plus sign, then the *constant* solution $X = T$, $Y = 1$ works:

$$\frac{dY}{dT} = \frac{-YT + \sqrt{1+T^2-Y^2}}{1+T^2} = \frac{-T + \sqrt{1+T^2-1}}{1+T^2} = \frac{-T+T}{1+T^2} = 0,$$

which is certainly true for Y a constant, and (again using the plus sign)

$$\frac{dX}{dT} = \frac{Y \pm T\sqrt{1+T^2-Y^2}}{1+T^2} = \frac{1+T\sqrt{1+T^2-1}}{1+T^2} = \frac{1+T^2}{1+T^2} = 1,$$

which is certainly true for $X = T$. This solution has a simple physical interpretation: $Y = 1$ means that $V_f y / b V_r = 1$, that is, y is a constant, and $X = T$ means that $V_f x / b V_r = V_f t / b$, that is, $x = V_r t$. But $y = \text{constant} \; (= c)$ and $x = V_r t$ is the mathematical description of the rabbit running along the straight-line path of figure 4.3.1, the special case we originally discussed.

The true value of this result is actually much more than just the solution to a special situation, because it sets limits on the behavior of *all possible* solutions (for the case where we use the plus sign with the radicals). This is because of the following deep and powerful theorem from the theory of differential equations: the first-order (perhaps even *nonlinear*) differential equation $dy/dx = f(x, y)$, with given initial conditions, has a *unique* solution. (You can find a proof of this in any good text on differential equations — look in the index for an entry on the *Picard solution* or *method*, named after the French mathematician Charles Picard (1856–1941)). This theorem tells us that no other possible solution to (4.3.8) and (4.3.10) can *cross* the constant solution, because the crossing point could then be taken as an initial condition and the "two" solutions would thereafter have to be identical by Picard's result.

So, we know not only that $Y(0) \leq 1$ but also that $Y(T) < 1$ for *all* $T > 0$ (in the case where we use the plus sign with the radicals). Thus, $0 \leq Y(T) < 1$ which says $Y^2(T) - 1 < 0$. And this, in turn, has a further implication. If you look at just the common radical term in (4.3.8) and (4.3.10) then you can see that we can write

$$\sqrt{1 + T^2 - Y^2} = \sqrt{Y^2 T^2 - (1 + T^2)(Y^2 - 1)} > YT,$$

where the inequality follows immediately because, as we just showed, $Y^2 - 1 < 0$. Thus, (4.3.8) says, *if we use the plus sign*, that

(4.3.12)
$$\frac{dY}{dT} > 0$$

and so $Y(T)$ is a monotonic increasing ("rising") function of T, bounded from above by 1, that is, $Y(T)$ increases constantly towards $Y = 1$. We can apply the same sort of argument to (4.3.10), that is, we can write

$$\frac{dX}{dT} = \frac{Y + T\sqrt{1+T^2-Y^2}}{1+T^2} = \frac{Y + T\sqrt{T^2-(Y^2-1)}}{1+T^2} > 0$$

because $Y^2 - 1 < 0$.

Next, what can we say about the *minus* sign (in front of each radical) case? From (4.3.8) we have, as before,

$$\frac{dY}{dT} = \frac{-YT - \sqrt{1+T^2-Y^2}}{1+T^2} = \frac{-YT - \sqrt{Y^2T^2-(1+T^2)(Y^2-1)}}{1+T^2},$$

or

(4.3.13) $$\frac{dY}{dT} < 0$$

because, as before, $Y^2 - 1 < 0$. That is, now Y is a monotonic *decreasing* ("falling") function of T. And as for dX/dT, from (4.3.10) we have

$$\frac{dX}{dT} = \frac{Y - T\sqrt{1+T^2-Y^2}}{1+T^2}$$

and now we have a new twist to consider; the possibility that $dX/dT = 0$, which by inspection occurs for $Y = T$. Furthermore, dX/dT can be either positive or negative, depending on Y: if $Y > T$ then $dX/dT > 0$, and if $Y < T$ then $dX/dT < 0$.

It may seem that we've learned a lot about the nature of the solutions to (4.3.8) and (4.3.10) without actually solving them, but there is actually still yet more we can deduce. For example, let's take another look at the possibility of $dX/dT = 0$. Since we can write, in that case,

$$\frac{dY}{dX} = \frac{dY/dT}{dX/dT} = \frac{\neq 0}{= 0} = \infty,$$

we see that the condition $dX/dT = 0$ means that the rabbit's hiding path curve $Y = Y(X)$ has a *vertical* tangent. This is, of course, geo-metrically clear when we observe that $dX/dT = 0$ is saying that X is

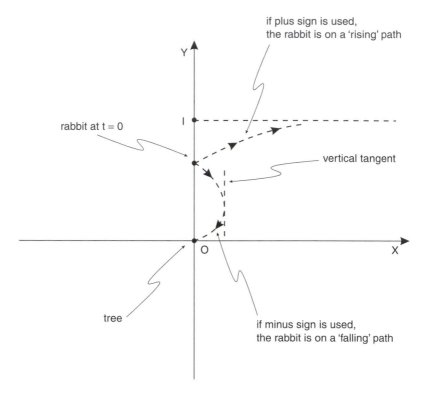

Figure 4.3.3 The hiding path curves for the plus and minus sign cases

not changing as T, and so Y, change, which can only happen if the $Y = Y(X)$ curve is vertical.

The value of all the prior discussion is that it allows us to sketch the general shape(s) of the rabbit's hiding paths in figure 4.3.3, even though we have not "solved" (4.3.8) and (4.3.10). For example, for the case of the minus sign in front of the radicals of (4.3.8) and (4.3.10), we know the hiding path must end up at the origin, that is, the rabbit eventually arrives at the tree. This is because of (4.3.9), which says $X = YT$, and so if $Y = 0$ then $X = 0$, too. And $Y = 0$ *must* occur (in the minus sign case) because we just showed that $dY/dT < 0$ in that case, that is, see (4.3.13). This "falling" path also clearly exhibits how a vertical tangent occurs in the minus sign case. Now, once the rabbit arrives at the tree she must make a decision. If she continues on a falling path she will no longer be hidden from the fox and so — if we

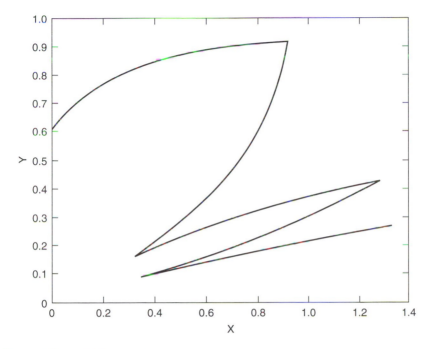

Figure 4.3.4 A hiding path

have an intelligent rabbit — she *won't* continue. Well, then, what *will* the rabbit do? She could simply stop running and sit motionless behind the tree (not mathematically very interesting and, in any case, not what we would expect a frightened rabbit to do with a nearby *moving* fox) or, if she does wish to continue running, she could switch to a "rising" path, that is, switch to using a plus sign with the radicals. Indeed, the rabbit could, *at any time*, switch signs and still remain hidden from the fox by the tree.

Figures 4.3.4 through 4.3.7 show four typical computer-generated[2] hiding paths for the rabbit. Figure 4.3.4 is for $Y(0) = 0.6$, an initial choice of the plus sign by the rabbit, and thereafter a switch in sign at integer values of T (all four plots are for $0 \leq T \leq 5$), while figure 4.3.5 assumes the initial sign choice is the minus sign. Figures 4.3.6 and 4.3.7 repeat the first two figures for a value of $Y(0) = 0.95$. The only exception to the sign changes at integer values of T is that the sign will also change if the hiding path becomes such that either $Y > 0.99$

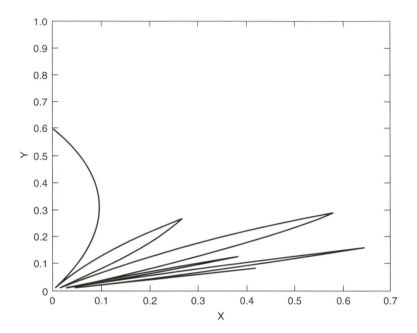

Figure 4.3.5 Another hiding path

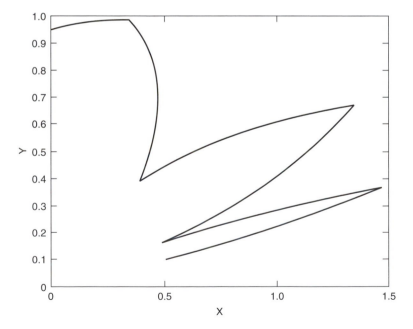

Figure 4.3.6 Yet another hiding path

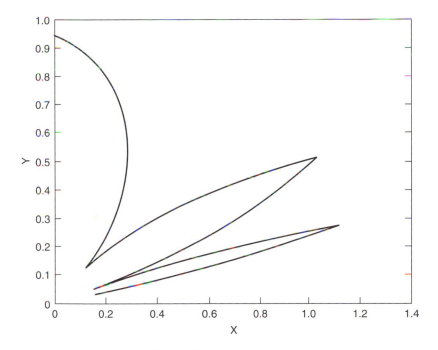

Figure 4.3.7 One more hiding path

or $Y < 0.01$, so as not to violate the double inequality on the value of $Y(T)$, that is, to be consistent with $0 \leq Y(T) \leq 1$. All four paths do, don't you think, look just like how you'd expect a skittish rabbit to run when frightened by a passing fox?

To end this section, try your hand at this challenge problem due to Professor Bailey. At the very end of Bailey (1994) we find the following remarks:

Polar coordinates (r,θ) are more natural for the hiding path problem and lead to the simpler differential equation

$$\frac{dR}{d\theta} = \pm\sqrt{\csc^4 \theta - R^2},$$

where $R = (rV_f)/(bV_r)$. However, the qualitative features of the solution are not as apparent from this equation as they were from [(4.3.8) and (4.3.10)]. The derivation of this polar equation provides an application of the arc length formula

$$(ds/d\theta)^2 = r^2 + (dr/d\theta)^2$$

in polar coordinates.

Your problem here is to derive Professor Bailey's differential equation. You'll find my derivation of it in appendix H.

4.4 The Hidden Object Problem: Pursuit and Evasion as a Simple Two-Person, Zero-Sum Game of Attack-and-Defend

In this section I want to take a break for just a bit from the heavy differential equation analysis we've been doing, and discuss a dramatically different technique. Here I'll show you how what is called *game theory* can have a role to play in pursuit and evasion analyses, *if* we are willing to loosen up a bit on what we mean by *pursuit* and *evasion*. In this section I'll replace those words with *attack* and *defend*, respectively. To start, I'll need to make some preliminary general comments to define our terminology.

A *two-person, zero-sum "game"* is a situation of complete conflict between two *rational* (I'll sharpen what this loaded word means mathematically soon) interests (or *players*), each of which is always in possession of all the information available to the other. Such a "game," for example, is the game of chess. (The reason I've put quote marks around *game* twice now is because the conflict situation may be very much *not* a game, e.g., it may be a military conflict of life-and-death.) Each player has multiple available *strategies* that may be employed when playing the game, but only one strategy may be used by each player during any given play; each player may, however, use a different strategy from one game play to the next. If a player does, in fact, use the same strategy over and over, to the exclusion of all other available strategies, then that player is said to be using a *pure* strategy. If, on the other hand, multiple strategies are used over a sequence of

		R #1	**R** #2
B	#1	6	5
B	#2	3	4

Figure 4.4.1 The pay-off matrix of an unfair game

games, then that player is said to be using a *mixed* strategy. A player can also play a mixed strategy even if the game is played only once, if the particular strategy actually used on the single play is selected by a probabilistic method. For example, if B has two possible strategies, and selects the one he will actually use with the flip of a fair coin, B is then said to be playing a mixed strategy with a probability distribution of $(\frac{1}{2}, \frac{1}{2})$. This interpretation of a mixed strategy will, in fact, be the one of central interest to us.

Each time the game of conflict is played, the outcome of the game — called the *game value* — is determined by examining what is called the *pay-off matrix* of the game. The pay-off matrix is simply a tabulation of what each player gains or loses given the two strategies that the two players decided to use on that particular game play. For the games we'll consider in this section, the gain of one player will be the loss of the other, hence the term *zero-sum*.

If, as is customary, we call the two players Blue (**B**) and Red (**R**), and if **B** has m available strategies from which to choose and **R** has n available strategies, then we'll call their conflict an $m \times n$ zero-sum game — in this section we'll be mostly interested in the case of $m = n = 2$. In such a game there will be a total of mn entries in the pay-off matrix. The usual practice is to enter a positive number if **B** wins for a given choice of strategies by **B** and **R**, and a negative number if **B** "loses" — which, since this is a zero-sum game, means **R** wins. For a 2×2 game with the pay-off matrix shown in figure 4.4.1, for example, the pay-off is 6 (to **B**) if both **B** and **R** play their strategies #1, while the pay-off is 5 (to **B**) if **B** and **R** play their strategies #1 and #2, respectively. Notice that this is an *unfair* game as, no matter what each player decides to do, **B** always wins because all the pay-off matrix entries are positive. One might wonder why **R** would play such a game; in fact, **R** may be *forced* to play, and I'll show you an example later in this section of a military problem with an all-positive entry pay-off matrix.

Now, let me draw a distinction between what it means in game theory to "choose a strategy" and what it means in the real world of, say, playing a game of chess. In real-world chess, the players — commonly called Black (**B**) and White (**W**), *not* Blue and Red! — alternate on making their moves: **W** moves first, **B** replies, **W** replies, etc., etc., etc. In game theory chess, each player writes out a *complete list* of all his moves, taking into account all possible replies the other player could make at each move. Such a list would require all the ink and paper that have been manufactured — and *will be* manufactured for the next one million years — actually to writeout. But remember, this is all theoretical and we are only to imagine such a list being prepared. Of course, each player could have *many* such enormous lists (one for every possible opening move, at least), with each *list* being a *strategy*. To play game theory chess, then, each player simultaneously chooses a particular list from his collection of lists and puts it on the table. One then "simply" cross-checks the two lists and sees whether checkmate or stalemate results.[3]

What this all has to do with us comes from a quest that has long intrigued mathematicians — the possibility of codifying human conflict in an analytic framework. The theory of probability (which found its origin in seventeenth-century analyses of games of chance, i.e., gambling) is one such highly successful development.[4] Most writers date the beginnings of *game theory* itself, as starting with the work of the French mathematician Emile Borel (1871–1956) in the early 1920s. It is in Borel's papers that one finds the concept of a mixed strategy, for example. Shortly after that, in 1928, came the seminal paper in what is called *modern* game theory with the work of the Hungarian-born American mathematician John von Neumann (1903–1957). In that paper he proved the central theorem of game theory, the so-called *minimax theorem*, which I'll discuss in just a bit. Later, in 1944, he coauthored the pioneering book *Theory of Games and Economic Behavior* with the Austrian economist Oskar Morgenstern (1902–1976), and it became an instant classic with enormous influence at the highest levels. The attack/defend problem I'm going to show you in this section, for example, is from Williams (1986) — a revision of a work originally published in 1954 — that represents the kind of two-person, zero-sum military game problems that followers of von Neumann studied at The RAND Corporation under the sponsorship of the U.S. Air Force.[5]

But, before I do show you that problem, there is one last issue we need to address: what does it mean when we say **B** and **R** are *rational*? Game theory is a *conservative* theory, and it defines rationality for **B** and **R** as follows: each will choose a strategy so as to do the best possible, *safely* (this is the crucial word!), under the assumption that the other player takes the same cautious decision-making approach. Specifically, in more detail,

> **B** chooses his strategy so that the *minimum* entry in the pay-off matrix, over all possible Blue strategies (the *rows* of the pay-off matrix), is *maximized*

and

> **R** chooses his strategy so that the *maximum* entry in the pay-off matrix, over all possible Red strategies (the *columns* of the pay-off matrix), is *minimized*.

Formally, what Blue does is pick strategy k such that row k in the pay-off matrix has the largest smallest element of all the rows, that is, if the entry in the i-*th* row ($1 \leq i \leq m$) and the j-*th* column ($1 \leq j \leq n$) of the pay-off matrix is a_{ij}, then Blue selects strategy k so that row k contains $\max\limits_{1 \leq i \leq m} \{\min\limits_{1 \leq j \leq n} (a_{ij})\}$. And what Red does is pick strategy l such that column l in the pay-off matrix has the smallest largest element of all the columns, that is, column l contains $\min\limits_{1 \leq j \leq n} \{\max\limits_{1 \leq i \leq m} (a_{ij})\}$. Now, *if* it happens to be the case that

$$\max_{1 \leq i \leq m} \{\min_{1 \leq j \leq n} (a_{ij})\} = \min_{1 \leq j \leq n} \{\max_{1 \leq i \leq m} (a_{ij})\} = V,$$

then we say that Blue's strategy k and Red's strategy l are *optimal* strategies. The reason we say this is because

  (a) if Blue selects strategy k then, *independent* of Red's choice, Blue is guaranteed of winning *at least V*;
  (b) if Red selects strategy l then, *independent* of Blue's choice, Red is guarenteed that Blue will win *no more* than V;
  (c) even if Red announced his strategy selection *ahead* of Blue's decision, Blue would still (if rational) choose his strategy k, and similarly Red would make the same choice of strategy even if Blue's choice were "leaked" ahead of time.

This interpretation of rational behavior has been called, by more than one wit, an example of "pray for the best but expect the worst" philosophy. That way, you'll never be disappointed and may even be pleasantly surprised! To see how it works in numbers, look again at the pay-off matrix of figure 4.4.1. If **B** plays his #1 strategy the least he wins is 5, while if **B** plays his #2 strategy the least he wins is 3. A rational **B**, therefore, picks his #1 strategy because $5 > 3$. For **R**, the reasoning goes as follows: if **R** plays his #1 strategy the most he loses is 6, while if he plays his #2 strategy the most he loses is 5. A rational **R**, therefore, plays his #2 strategy because $5 < 6$. The value of this game is 5 and, it is most important to note, *there is no incentive for either player to change his decision on subsequent plays of the game* (that is, **B** and **R** — if rational — should each play a pure strategy). If **B** did, for some reason, switch to his #2 strategy he would then win at most 4 instead of 5 (and **R** could force **B**'s winnings farther down to 3 by switching to his #1 strategy), and if **R** did, for some reason, switch to his #1 strategy he could then lose 6 (if **B** then switched to his strategy #1) instead of only 5.

When a game analysis leads to this sort of result, with the maximum over the minimum of the pay-off matrix row entries equal to the minimum over the maximum of the pay-off matrix column entries, we have a *stable* game, that is, a game in which there is no incentive for either player to change strategies from one play to the next. Such a game is said to have a *saddle point*. This picturesque term comes from the geometry of a saddle — skipping a little bit out of sequence, take a look-ahead-peek at figure 4.4.3 — which curves downward in the

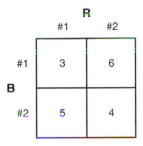

Figure 4.4.2 A two-person, 2 × 2, zero-sum game without a saddle-point

front-to-back direction, and upward in the side-to-side direction. In the middle of the saddle, therefore, we have a point, the *saddle point*, that is a local *minimum* in one direction (front-to-back) and simultaneously a local *maximum* in the other direction (side-to-side). Not all games have a saddle point, however; consider, for example, the pay-off matrix shown in figure 4.4.2. To maximize his minimum gain, **B** would choose his #2 strategy (4 > 3), while to minimize the maximum gain for **B** (i.e., to minimize the maximum entry in the columns of the pay-off matrix), **R** would chose his #1 strategy (5 < 6). Notice that now **B**'s maximum over the minimums of the rows does *not equal* **R**'s minimum over the maximum of the columns (of the pay-off matrix).

The result, unlike that in our first example, is that now there *is* incentive for both players to change strategies. For example, **R** changing from his #1 strategy to his #2 strategy reduces the pay-off to **B** from 5 to 4. But then there is incentive for **B** to change from his #2 strategy to his #1 strategy because that increases the pay-off (to himself) from 4 to 6. And so on, *ad infinitum*. For either **B** or **R** to play a pure strategy would be a mistake — rather, each should play this game with a *mix* of strategies. It is obviously important for each player, however, to ensure that the other player does not have advance information on what particular strategy will be played against him. Hence the use by both **B** and **R** of *random* mechanisms (with associated probability distributions) to decide the strategy to be employed by each for any given play of the game. That way, there can be no advance information "leak" in either direction because even **B** and **R** themselves don't have advance information. Neither knows what they'll do until they actually do it!

It was von Neumann's great achievement, in his 1928 paper, to prove that for *any* two-person, zero-sum game there exist probability distributions for mixed strategies, for both **B** and **R**, such that the average pay-off for both players is the same value V and, in addition, that V is the best each can expect to achieve. This is von Neumann's celebrated *minimax theorem*, which I'll not prove here as you can find a proof of it in just about any good technical book on game theory. (Saddle-point games, with pure strategies, are simply special cases of games with mixed strategies that have probability distributions with all the strategy probabilities equal to zero except for one, and that one strategy has probability one.)

To see how this works in numbers, let's workout the probability distributions for the game of figure 4.4.2. Let's suppose **B** plays his #1 strategy with probability p, and his #2 strategy with (of course) probabilty $1 - p$. Similarly, **R** plays his #1 strategy with probability q, and his #2 strategy with (of course) probabilty $1 - q$. If **B** plays his strategy #1 then he wins 3 with probability q and 6 with probability $1 - q$, for an average winnings of $3q + 6(1 - q)$. Similarly, if **B** plays his strategy #2 he wins 5 with probability q and 4 with probability $1 - q$, for an average winnings of $5q + 4(1 - q)$. Since **B** plays his strategy #1 with probability p and his strategy #2 with probability $1 - p$, **B**'s overall average winnings are

$$V(p, q) = p[3q + 6(1 - q)] + (1 - p)[5q + 4(1 - q) = 4 + 2p + q - 4pq.$$

Of course, we could just as well have looked at this calculation from **R**'s point of view. If **R** plays his strategy #1 then **B** wins 3 with probability p and 5 with probability $1 - p$, for an average winnings of $3p + 5(1 - p)$. Similarly, if **R** plays his strategy #2 then **B** wins 6 with probability p and 4 with probability $1 - p$, for an average winnings of $6p + 4(1 - p)$. Since **R** plays his strategy #1 with probability q and his strategy #2 with probability $1 - q$, then **B**'s overall average winnings are

$$q[3p + 5(1 - p)] + (1 - q)[6p + 4(1 - p)] = 4 + 2p + q - 4pq = V(p, q),$$

just as computed before. Figure 4.4.3 shows a three-dimensional plot of $V(p, q)$, which makes it quite clear that there is indeed a minimax saddle point to this game played with mixed strategies.

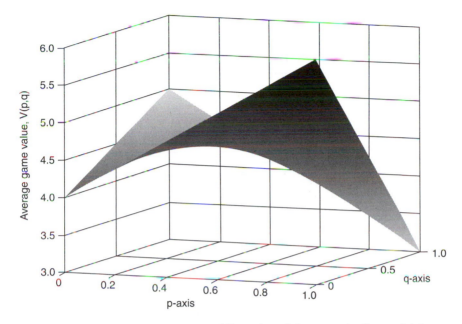

Figure 4.4.3 Seeing the minimax saddle-point of the game in figure 4.4.2

Now, **B** wants V to be as large as he can safely get it, while **R** wants V to be as *small* as he can safely get it. Looking at the game from **B**'s point of view, suppose he has set his value for p. To do the best he safely can *on average*, he wants his p to be such that the game is at its mixed strategy saddle-point, which means that

$$\frac{\partial V}{\partial q} = 0,$$

that is, **B** wants to pick p such that $1 - 4p = 0$. Thus, $p = \frac{1}{4}$ and so **B**'s optimal probability distribution is $(\frac{1}{4}, \frac{3}{4})$. Let's now look at the game from **R**'s point of view. Suppose he has set his value for q. To do the best he safely can *on average*, **R** wants his q to be such that the game is at its mixed strategy saddle point, which means that

$$\frac{\partial V}{\partial p} = 0,$$

that is, he wants to pick q such that $2 - 4q = 0$. Thus, $q = \frac{1}{2}$ and so **R**'s optimal probability distribution is $(\frac{1}{2}, \frac{1}{2})$. The average value of the game (the saddle-point value) is, as shown in Figure 4.4.3,

$$V\left(p = \frac{1}{4}, q = \frac{1}{2}\right) = 4 + 2\left(\frac{1}{4}\right) + \frac{1}{2} - 4\left(\frac{1}{4}\right)\left(\frac{1}{2}\right) = 4 + \frac{1}{2} + \frac{1}{2} - \frac{1}{2} = 4.5.$$

Now, at last, we come to the problem I want to show you, a problem presented in Williams (1986) as follows, in the context of a military conflict between three aircraft, a fighter and two bombers:

> Suppose that a pair of Blue bombers is on a mission; one carries the bomb [this scenario was created during the 1950s Cold War between the Soviet Union and America and "the bomb" meant the hydrogen bomb] and the other carries equipment for radar jamming, bomb damage assessment, or what-have-you. These bombers fly in such a way that Bomber 1 derives considerably more protection from the guns of Bomber 2 than Bomber 2 derives from those of Bomber 1. There is some concern lest isolated attacks by one-pass Red fighters [an accidental but oh-so-Cold-War suggestion of *Soviet* fighters] shoot down the bomb carrier, and the survival of the bomb carrier transcends in importance all other considerations. The problem is: Should Bomber 1 or Bomber 2 be the bomb carrier, and which bomber should the Red fighter attack?

This is a problem in which the **R** fighter *pursues* the bomb, and **B** attempts to *evade* (to hide the bomb from) the fighter.

We can list the available strategies for both **B** and **R** as follows:

B strategy #1: carry bomb in Bomber 1 (more protected bomber)
B strategy #2: carry bomb in Bomber 2 (less protected bomber)
R strategy #1: attack Bomber 1
R strategy #2: attack Bomber 2

Probably most people, if presented with this statement of the problem but with no knowledge of game theory, would intuitively say that the bomb should *always* be placed on the more protected bomber (Bomber 1) and, therefore, the fighter should *always* go after Bomber 1. As you'll soon see, however, while that may be intuitive it is also wrong. Continuing with the specifics given in Williams (1986),

R
#1   #2

	#1	80	100
B			
	#2	100	60

Figure 4.4.4 The fighter vs. bombers pay-off matrix

"Suppose the chance the bomb carrier will survive, if attacked, is 60 percent in the less [protected bomber] and 80 percent in the more [protected bomber] and is 100 percent if it is not attacked". With these numbers Williams (1986) writes the pay-off matrix of this game as shown in figure 4.4.4.

The fighter versus the bombers "game" clearly has no pure strategy saddle-point, as the maximum of the row minimums is 80, while the minimum of the column maximums is 100. So, we need to do a mixed strategy analysis. Notice, too, that with an all-positive entry pay-off matrix this "game" is unfair to **R**, but of course the concept of *fairness* has no meaning in a military conflict; **R** *must* play the "game" as his only alternative is simply to let **B**'s bombers attack without meeting any resistance at all. Even with an all-positive entry pay-off matrix, of course, **R** could still actually *win* by successfully shooting down the bomb carrier — the physical significance of the value of the game is as the *probability* that **B**'s bomb carrier will survive, and that is not a sure thing. So, even before we do any analysis it should be clear that the mixed strategy saddle-point value of the game will be less than 100.

Let **B** play his strategies with the probability distribution $(p, 1 - p)$ and **R** play his strategies with the probability distribution $(q, 1 - q)$. Then, with $V(p, q)$ as the average game value, we can write it as either

$$V(p, q) = p[80q + 100(1 - q)] + (1 - p)[100q + 60(1 - q)],$$
$$V(p, q) = q[80p + 100(1 - p)] + (1 - q)[100p + 60(1 - p)].$$

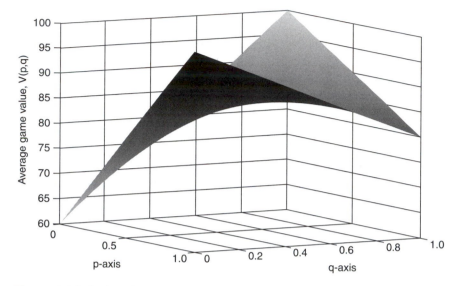

Figure 4.4.5 Seeing the minimax saddle-point of the fighter vs. bombers game

You should verify that both of these expressions do in fact reduce to the same result:

$$V(p, q) = 60 + 40p + 40q - 60pq.$$

Figure 4.4.5 shows the mixed strategy minimax saddle point of this game.

At the minimax point we have $\partial V/\partial p = \partial V/\partial q = 0$, which says $40 - 60q = 0$ and $40 - 60p = 0$ or, $p = q = \frac{2}{3}$. That is, the optimal mixed strategies for **B** and **R** have the same probability distribution $(\frac{2}{3}, \frac{1}{3})$. So, contrary to intuition, with probability $\frac{1}{3}$ **B** should place the bomb on the *less* protected bomber, and with probability $\frac{1}{3}$ **R** should attack that bomber. With these optimal strategies the game value at the minimax point is

$$V\left(\frac{2}{3}, \frac{1}{3}\right) = 60 + 40\left(\frac{2}{3}\right) + 40\left(\frac{2}{3}\right) - 60\left(\frac{2}{3}\right)\left(\frac{2}{3}\right) = 60 + \frac{160}{3} - \frac{80}{3}$$

$$= 60 + \frac{80}{3} = 86\frac{2}{3},$$

that is, the bomb carrier will survive with probability 0.8667, which is greater than the survival probability of 0.8 that **B** would achieve if he just did the "obvious thing" and *always* placed the bomb on the more protected bomber.

What if **B** knows game theory but **R** doesn't? That is, $p = \frac{2}{3}$ but q is *not* necessarily $\frac{2}{3}$? Then,

$$V\left(\frac{2}{3}, q\right) = 60 + 40\left(\frac{2}{3}\right) + 40q - 60\left(\frac{2}{3}\right)q = 86\frac{2}{3} + 40q - 40q = 86\frac{2}{3},$$

a result *independent* of q. That is, if **B** plays his optimal strategy then it *doesn't matter* what **R** does! **B**'s optimal play *ensures* a game value of $86\frac{2}{3}$ *independent* of **R**'s decision-making process. Similarly, if $q = \frac{2}{3}$ but p is not necessarily $\frac{2}{3}$, then

$$V\left(p, \frac{2}{3}\right) = 60 + 40p + 40\left(\frac{2}{3}\right) - 60p\left(\frac{2}{3}\right) = 60 + 40p$$

$$+ \frac{80}{3} - 40p = 86\frac{2}{3},$$

a result *independent* of p. So, if **R** plays his optimal strategy then it *doesn't matter* what **B** does: **R**'s optimal play ensures a game value of $86\frac{2}{3}$ *independent* of **B**'s decision-making process. Amazing.

Now, here's a challenge problem for you, yet another military conflict situation from Williams (1986), the so-called "Attack-Defense Game":

> Blue has two installations. He is capable of successfully defending either of them, but not both; and Red is capable of attacking either, but not both. Further, one of the installations is three times as valuable as the other. What strategies should they adopt?

As a hint, here's what Williams wrote next:

> Take the value of the lesser installation to be 1. Then, if both survive the payoff is 4; if the greater survives the payoff is 3; if the lesser survives the payoff is 1. Designating the defense (or attack) of the lesser installation as Strategy 1, we have [the pay-off matrix of figure 4.4.6]:

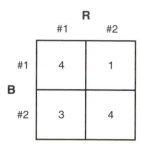

Figure 4.4.6 The pay-off matrix of the attack-defense game

This game does not have a pure strategy saddle point, as the maximum of the row minimums (3) does not equal the minimum of the column maximums (4). Thus, you need to do a mixed strategy analysis. That is, you are to determine the optimal probability distributions for **B** and **R**. You can check your answers against mine in appendix I. Are you surprised by the answers?

4.5 The Discrete Search Game for a Stationary Evader — Hunting for Hiding Submarines

With the introduction in the previous section of game-theoretic ideas into pursuit-and-evasion, let me elaborate just a bit more along the same *general* lines, but in terms of an entirely different appearing problem. The problem I'll describe here will be somewhat abstract, but it could serve as a first-order model for the problem of hunting for a submerged submarine that is hiding, *motionless*, at any one of a number of possible locations on the floor of the ocean.[6] To quote Ruckle (1991),

> A Hider hides in one of the locations 1, 2, 3, ..., n [and remains there]. The Seeker searches the locations, one after the other with returns allowed. He knows that if he visits location i, he will find the Hider there with chance $0 < p_i \leq 1$, provided the Hider is there.... The pay-off to the Hider is the number of searches the Seeker must make before locating him. [The case $p_i = 0$ is excluded since that means hiding place i is *perfect* (from the point of view of the Hider); the strategy of the Hider is trivially obvious.]

So, here is a game in which the Hider wishes to maximize the number of searches, while of course the Seeker has precisely the opposite goal of minimizing the number of searches. You can now surely see what our general questions are: what is the "optimal" strategy for the Hider to employ in selecting a hiding place, and what is the "optimal" strategy for the Seeker to employ in searching for that hiding place?

Before going farther, notice that the problem we are going to examine states that the submarine (the evader) is not moving. If we allow for the possibility of the evader to move from one hiding place to another while the search is going on, then we will have a *much* more difficult problem — see, for example, Dobbie (1974). Such problems do occur, of course (our motionless evader is, after all, simply the special case of a moving evader with zero speed!). The coast guards of all maritime nations are faced with just such a problem when searching for a lost ship that is being moved around the ocean surface *at random* by both wave and wind actions — see, for example, Mangel (1981) and Stone (1977). Problems of *intelligent* moving evaders are discussed in Washburn (1980), Eagle (1984), and Baston and Bostock (1989). I'll discuss one example of the complications introduced by a mobile evader in the next section. But, for now, back to our stationary evader problem.

There is a very technical theory with which to attempt to derive the optimal strategies for both the Hider and the Seeker — which will turn out to be, in general, mixed strategies — but I will limit this section to two very special (but also very important) cases, and to giving their solutions without proofs. The proofs to both cases can be found in Ruckle (1991). To start, we need to define what it means to talk of a *search strategy* for the Seeker. Most simply, a search strategy is an *ordered list* of the hiding locations, in the order with which the Seeker will search them until the Hider is found. Thus a strategy list is in general infinitely long, since there is a nonzero probability, with every look at a hiding location, that the Hider will not be detected *even if there*. To put this more formally, let's make the following definition. $f_j(i)$ will denote the *j*th strategy of the Seeker, that is, the Seeker's *j*th ordered list, where $f_j(i) = l$ means that for the *j*th strategy the Seeker searches hiding location l at time i, $i = 1, 2, 3, \ldots$ and $l = 1, 2, 3, \ldots, n$. $f_j(i)$ is called a *search sequence function*.

We can now state our first special case. If all the p_i are equal (equal to p), then:

(a) an optimal strategy for the Hider is to hide at location l with probability $\frac{1}{n}$ for location $l = 1, 2, 3, \ldots, n$;

(b) an optimal strategy for the Seeker is to choose one of the following search sequence functions with probability $1/n$:

$$f_j(i) = 1 + (i + j)\mathrm{mod}\ n, \ j = 1, 2, 3, \ldots, n; i = 1, 2, 3, \ldots;$$

(c) the average value of the game (the average duration of the search) is $\frac{n}{p} - (n-1)/2$ looks.

For example, if $n = 3$ and $p = 0.3$, then

(a) an optimal strategy for the Hider is to hide at location l with probability $\frac{1}{3}$ for location $l = 1, 2, 3$;

(b) an optimal strategy for the Seeker is to choose one of the following search sequence functions with probability $\frac{1}{3}$:

$$f_1(i) = 1 + (i + 1)\mathrm{mod}\ 3 = 3, 1, 2, 3, 1, 2, \ldots,$$

$$f_2(i) = 1 + (i + 2)\mathrm{mod}\ 3 = 1, 2, 3, 1, 2, 3, \ldots,$$

$$f_3(i) = 1 + (i + 3)\mathrm{mod}\ 3 = 2, 3, 1, 2, 3, 1, \ldots;$$

(c) the average duration of the search is $3/0.3 - (3-1)/2 = 10 - 1 = 9$ looks.

The result in (a) makes intuitive sense, by symmetry, but the result in (c) is not at all obvious. We can "check" it easily enough, however, or at least verify that it is consistent with (a) and (b), by writing a computer program incorporating a random number generator (a so-called "Monte Carlo" program) to simulate the seach-and-hide processes. This approach also has the added feature of giving us estimations

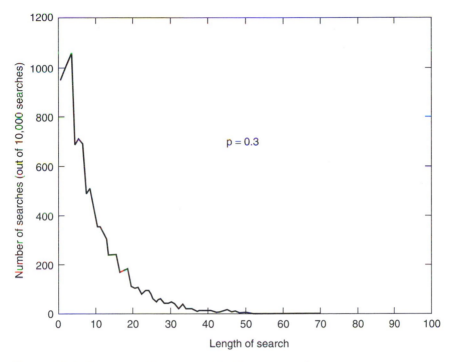

Figure 4.5.1 Computer simulation of a discrete search

of the likelihood of search durations deviating significantly from the average value. When I wrote and ran such a simulation (for $p = 0.3$), for example, simulating $10,000$ searches, the result was an average duration of 9.0726 (within 1% of the theoretical value). To show the variations of the search duration around the average value, the simulation kept track of the duration of each of the $10,000$ searches and generated figure 4.5.1. The longest search had a (surprisingly long, I think) duration of 83.

For our second special case, I'll now restrict the number of hiding places to just two, one of which (location 1) has two special characteristics. It is simultaneously more difficult for the Hider to *reach* to hide in than is the other hiding place (location 2) and, *if* examined by the Seeker, it is very *easy* to search. Because of this second characteristic it is not likely that the Hider will go to location 1 — although he *might* go there *just because* he knows the Seeker thinks it *unlikely*.

(Oh, my, the convoluted thinking *that* prompts!) To model this, we'll take the probability of detecting the Hider in location 1 as 1 (*if* the Hider is, fact there), while the probability of detecting the Hider in location 2 is p (*if* the Hider is there, of course). So, for this case we have $n = 2$, $p_1 = 1$, and $p_2 = p$. As before, I'll simply state the optimal strategies for the Hider and the Seeker, and refer you to Ruckle (1991) for a proof.

Perhaps surprisingly, the strategies for this $n = 2$ case are just a bit more complicated than the strategies in the first case we considered, which was for *arbitrary n*.

(a) An optimal strategy for the Hider is first to determine the value of the largest integer h less than

$$\frac{1}{p} + (1 - p)^{h-1},$$

then to calculate the value of

$$\alpha = \frac{1}{1 + p(1 - p)^{h-1}},$$

and then to hide at location 1 with probability $1 - \alpha$ and at location 2 with probability α.

(b) An optimal strategy for the Seeker is first to determine the value of

$$a = \frac{h + 1 - 1/p - (1 - p)^h}{1 + p(1 - p)^{h-1}},$$

and then to use the search sequence function $f_h(i)$ with probability a, and the search sequence $f_{h+1}(i)$ with probability $1 - a$, where the search sequence function f_m has the values of $f_m(m) = 1$ and $f_m(i) = 2$ for $i \neq m$.

(c) The average search duration is $h + 1 - a$ looks.

For example, with $p = 0.3$ we have

(a) To find h, we construct the following table:

h	$1/0.3 + (0.7)^{h-1}$
1	4.3333
2	4.0333
3	3.8233
4	3.6763

and so we see that $h = 3$. Then

$$\alpha = \frac{1}{1 + (0.3)(0.7)^2} = 0.872,$$

and so the Hider should hide at location 1 with probability 0.128 and at location 2 with probability 0.872.

(b) Thus,

$$a = \frac{3 + 1 - 1/0.3 - (0.7)^3}{1 + (0.3)(0.7)^2} = \frac{0.324}{1.147} = 0.282,$$

and so the Seeker should use search sequence function $f_3(i) = 2, 2, 1, 2, 2, 2, 2, \ldots$ with probability 0.282 and search sequence function $f_4(i) = 2, 2, 2, 1, 2, 2, 2, \ldots$ with probability 0.718.

(c) The average search duration is $3 + 1 - 0.282 = 3.718$ looks.

As with the first special case we considered, a Monte Carlo simulation of this case is not difficult to write. The result of a 10,000-search simulation, with $p = 0.3$, is shown in figure 4.5.2, with an average search duration of 3.7613 looks (again about 1% off from the theoretical value). The longest search had a duration of 29 looks.

This section is only a tiny scratch on the surface of an iceberg of search analyses that have been published, and continue to be published. Open any issue of such journals as *Operations Research*, *Journal of Optimization Theory and Applications*, and *Naval Research Logistics*, and the odds are pretty good it will have a paper (or two or three) on search theory.

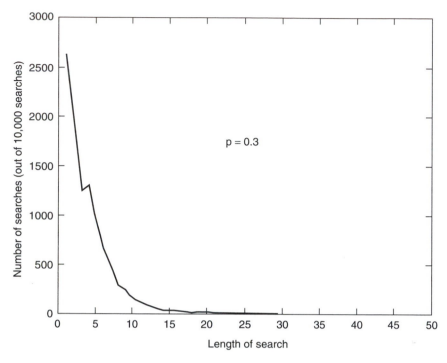

Figure 4.5.2 Computer simulation of a discrete search

4.6 A Discrete Search Game with a Mobile Evader — Isaacs's Princess-and-Monster Problem

In this section I'll show you a classic problem of pursuit-and-evasion involving a *mobile* evader. You are to imagine that the chase takes place in a pitch-black circular tunnel: neither the pursuer nor the evader — they move at the same speed — can see the other, and capture occurs as a surprise to both when (if) they collide as they travel through the tunnel. As you might expect, the pursuer wants capture to occur as soon as possible, while the evader wishes to delay capture for as long as possible. Our question is as before: what are the optimal strategies for the pursuer and the evader?

This game is commonly called the Princess-and-Monster Game (the pursuer is obviously the monster!) and it is due to the American mathematician Rufus Isaacs, whose guarding-the-target problem was the topic of section 4.2. What I'll discuss here is the solution to the discrete space-and-time version of Isaacs's problem.[7] As in the previous section on stationary evaders, I'll just describe the optimal strategies for both the monster and the princess, do a partial validation of those strategies with a Monte Carlo simulation, and refer you to the literature — in this case Wilson (1972) — for the proofs. Before I can do that, however, I need to make some definitions and establish some notation.

We'll take our tunnel as represented by $n \geq 3$ position points uniformly distributed along the length of the tunnel (this is the *spatial* discrete part of the problem). During the monster's (M) pursuit of the princess (P), which starts at time $t = t_0 = 0$ with both M and P at specified initial position points — which both M and P know for themselves *and* for the other — time will proceed as *tick-tock time*, in a regular, discrete manner, as $t = t_1, t_2, t_3, \ldots$. M and P are, at every instant of tick-tock time, at one of the n position points along the tunnel. At each new tick-tock instant, t_k, both M and P decide to move either *one* position point clockwise, or *one* position point counterclockwise, to arrive at their position point for time t_{k+1}. Both M and P always know what *they* have done, but neither has any knowledge of what the other has done. The chase proceeds in tick-tock time in this way until (if) one of two things occurs:

(a) after M and P have each made their last move they are at the same position point;
(b) after M and P have each made their last move they have *interchanged* their position points from those at the previous time.

If either (a) or (b) occurs at time $t = t_k$ then we say M has captured P at time $t = t_k$. When necessary for computational purposes, we'll take $t_k = k$.

To understand what M and P should each do to accomplish their respective goals, we need to define what is called the *initial state* of the chase, and what are called *escape sequences*. The initial state is the easier of the two concepts. With reference to figure 4.6.1, we mark off the n position points along our tunnel where, without loss of generality, we can always assume M is at position point 0 at time $t = t_0 = 0$, on the

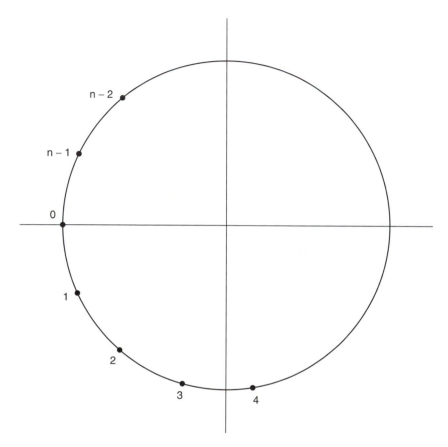

Figure 4.6.1 The n position points along the tunnel, with the monster at position 0 at time $t = t_0 = 0$

horizontal axis to the far left. If we take P's initial position point to be j ($1 \leq j \leq n-1$), then we say the initial state of the pursuit is j (the state of the pursuit, in general, is P's position point value minus M's position point value). Next, with our values of n and j in hand, we defined a third parameter, v, as follows:

(4.6.1) $v = \begin{cases} \left\lceil \dfrac{n}{2} \right\rceil & \text{when } j \text{ and } n \text{ are not both even,} \\[2ex] \left(\dfrac{n}{2}\right) - 1 & \text{when } j \text{ and } n \text{ are both even.} \end{cases}$

(The notation $[x]$ means the integer part of x, e.g., $[6.1] = 6$.) We'll find that v plays a central role in determining the duration of the pursuit.

With the values of n and j in hand, we can now define what are called *escape sequences*, a term I'll explain at the end of this section. In what follows in these definitions, you'll see strings of $\pm 1s$, which have the following physical interpretations:

$+1 \Rightarrow$ move one position point counterclockwise
$-1 \Rightarrow$ move one position point clockwise.

So,

(ES1) If j is odd and n is even, the escape sequence is

$$-1, -1, -1, \ldots, -1 \text{ for } t = t_k, \ 1 \leq k \leq \frac{j-1}{2},$$

$$+1, +1, +1, \ldots, +1 \text{ for } t = t_k, \ \frac{j+1}{2} \leq k \leq \left(\frac{n}{2}\right) - 1.$$

(ES2) If j is odd and n is odd, the escape sequence is

$$-1, -1, -1, \ldots, -1 \text{ for } t = t_k, \ 1 \leq k \leq \frac{j-1}{2},$$

$$+1, +1, +1, \ldots, +1 \text{ for } t = t_k, \ \frac{j+1}{2} \leq k \leq \frac{n-1}{2} - 1.$$

(ES3) If j is even and n is odd, the escape sequence is

$$-1, -1, -1, \ldots, -1 \text{ for } t = t_k, \ 1 \leq k \leq \left(\frac{j}{2}\right) - 1,$$

$$+1, +1, +1, \ldots, +1 \text{ for } t = t_k, \ \frac{j}{2} \leq k \leq \frac{n-1}{2} - 1.$$

(ES4) If j is even and n is even, the escape sequence is

$$-1, -1, -1, \ldots, -1 \text{ for } t = t_k, \ 1 \leq k \leq \left(\frac{j}{2}\right) - 1,$$

$$+1, +1, +1, \ldots, +1 \text{ for } t = t_k, \ \frac{j}{2} \leq k \leq \left(\frac{n}{2}\right) - 2.$$

With these definitions completed, I can now give you the optimal strategies for both the monster and the princess. Quoting Wilson (1972), where the reference to a coin is to a *fair* coin (equal probabilities of a head and a tail), and v is as defined in (4.6.1),

> The optimal strategies... have a simple interpretation in terms of coin tossing. [The monster] can play [his strategy] by tossing a coin prior to his 1st, $(v+1)$st, ..., $(mv+1)$st moves and then moving clockwise for the succeeding v moves if the coin shows heads and counterclockwise if it shows tails. [The princess can play her strategy] by using the [appropriate escape sequence, as determined by n and j] for [her] first $v-1$ moves, which will guarantee [her] safety during these moves and bring [her] on [her] $(v-1)$st move to a point which is as close as possible to the antipode of [the monster's] starting point. [The princess] then plays similarly to [the monster], tossing a coin prior to [her] vth, $2v$th, ..., mvth moves and moving clockwise or counterclockwise for the succeeding v moves according to whether the coin shows heads or tails.

The average duration of the pursuit is given by

$$\sum_{q=1}^{\infty} \frac{t_{qv}}{2^q} = \frac{t_v}{2} + \frac{t_{2v}}{4} + \frac{t_{3v}}{8} + \frac{t_{4v}}{16} + \cdots,$$

or, since $t_k = k$, the average duration of the pursuit is given by

(4.6.2) $$\frac{v}{2} + \frac{2v}{4} + \frac{3v}{8} + \frac{4v}{16} + \cdots,$$

an infinite series easily summed to $2v$. Notice that this tells us the average duration for *any* pursuit, no matter what n and j may be, will *always* be an *even integer*.

To see how this procedure works in numbers, suppose $n = 10$. If we start P at the furthest she can be from M, she'll be at position location 5 (and by convention M is at position location 0). That is, $j = 5$. This is illustrated in figure 4.6.2. And so, since j and n are not both even we

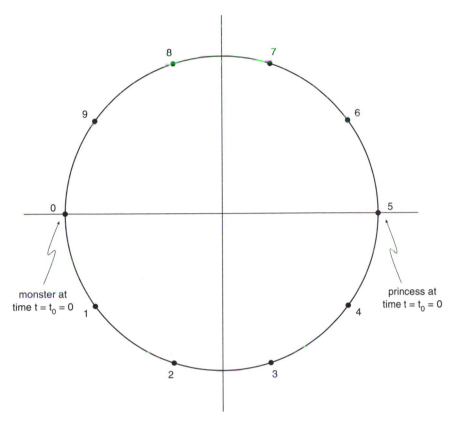

Figure 4.6.2 The initial state of the Princess-and-Monster pursuit for $n = 10$, $j = 5$, and $v = 5$

have, by (4.6.1), that $v = [\frac{10}{2}] = [5] = 5$. M therefore can precalculate his moves as follows:

(a) With probability $\frac{1}{2}$ he will move clockwise for the first five moves, and with probability $\frac{1}{2}$ he will move counterclockwise for the first five moves.

(b) If, assuming capture has not occurred, then with probability $\frac{1}{2}$ he will move clockwise for the next five moves, and with probability $\frac{1}{2}$ he will move counterclockwise for the next five moves.

(c) Repeat (b).

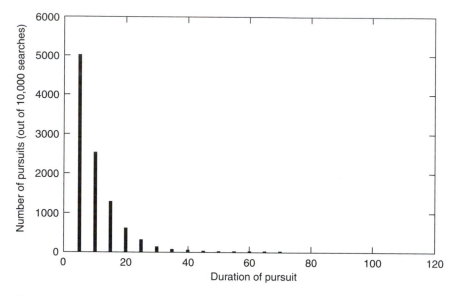

Figure 4.6.3 Computer simulation of a Princess-and-Monster Pursuit

P can precalculate her moves as follows:

(a) Since $n = 10$ and $j = 5$, i.e., n is even and j is odd, she will start by using escape sequence ES1, i.e., she will move clockwise on her first two moves and then counterclockwise on her next two moves.

(b) If, assuming capture has not occurred, then with probability $\frac{1}{2}$ she will move clockwise for the next five moves, and with probability $\frac{1}{2}$ she will move counterclockwise for the next five moves.

(c) Repeat b.

Notice that no matter the direction of moving for either M or P, it is *impossible* for capture to occur before the fifth move. This, in fact, is the motivation for the term *escape sequence* for P's initial moves — whatever the values for n and j, the associated escape sequence ensures that P can escape capture for at least some minimum time. And finally, for this pursuit example, the average duration is $2v = 2(5) = 10$.

This is all very easy to code in a Monte Carlo simulation and, when I did that for 10,000 pursuits, the average duration was 9.9735 moves (a less than 0.3% error). Figure 4.6.3 shows the variation in pursuit durations around this average value, with the longest pursuit lasting for 70 moves. Notice, too, that the shortest pursuits did, indeed, as explained, have durations equal to five moves. The simulation hints strongly at the conclusion that the pursuit duration of any *individual* pursuit in this particular example will always be a multiple of five.[8]

4.7 Rado's Lion-and-Man Problem and Besicovitch's Astonishing Solution

In the final, brief section of this chapter I want to show you a celebrated pursuit-and-evasion problem that, for a while, led *modern* mathematicians astray. It was created by the German-born mathematician Richard Rado (1906–1989) who worked all his professional life in England after escaping from the Nazis in 1933. Sometime in the mid-1920s Rado posed the following apparently simple question (he never actually *published* his problem, but rather it spread by word-of-mouth): A lion and a man are both inside a circular arena, each running at the same constant speed. Can the lion catch the man? We are not to think of the lion and man as spatially extended objects, but rather as points.

After Rado moved to England in 1933 — where he earned his second of two doctorates in mathematics under the direction of the famous English mathematician G. H. Hardy (1877–1947) — he became friendly with Hardy's long-time collaborator John Littlewood (1885–1977), who wrote of Rado's problem in his 1953 book *A Mathematician's Miscellany* (reprinted in 1986 as *Littlewood's Miscellany*) as follows:

> It was said that the "weighing-pennies" problem wasted 10,000 scientist-hours of war-work, and that there was a proposal to drop it over Germany. This one, though 25 years old, has recently swept the country; but most of us were teased no more than enough to appreciate a happy idea before arriving at the answer [to Rado's problem], "L keeps on the radius OM."

That is, if O is the center of the arena, the lion should — according to this answer — first get on a radius connecting O to M and thereafter keep itself, the man, and O collinear. Croft (1964) calls this the "radius rule." Does this strategy sound familiar to you? It should, as it appears to be an only modestly disguised version of Houghton's problem that we discussed in section 2.4! You'll soon see, however, that the resemblance is only superficial.

Of course, the lion-and-man problem *is* Houghton's problem *if* the man actually does run along a circular path centered on the center of the arena. But what if he *doesn't* do that? There is nothing in the problem statement that says he has to — indeed, as Littlewood went on to write in his book, "the 'answer' is wrong, and M *can* [my italics] escape capture... This has just been discovered by Professor A. S. Besicovitch." Littlewood was referring to the Russian-born mathematician Abram Besicovitch (1891–1970), who had emigrated to England in the mid-1920s, where he met and worked with both Littlewood and Rado. Decades after Rado first posed his problem, Besicovitch showed that there is a (polygonal) outward spiraling path that the man can run along that will allow him to avoid capture by the lion — which we assume is following the "radius rule" — for an arbitrarily long time (math-speak for the man to never be captured). The lion will get arbitrarily close to the man, yes, but since we are taking both the lion and the man as *point* objects capture never occurs. This nonintuitive result astonished the mathematical world and yet, as you'll soon see, it is not at all difficult to describe Besicovitch's path — to *discover* it, of course, required Besicovitch's genius!

To demonstrate that the long-believed "radius rule" strategy for the lion fails to ensure capture, we can make whatever special assumptions we wish about what the man does — all we need for a refutation is just *one* counterexample. So, let's take the lion, at $t = 0$, at the center of a circular arena with radius one, and the man as initially at $(0, \frac{1}{2})$. We next imagine that the pursuit motion of the lion, and the evasive action of the man, are represented by a *discrete-time, turn-based* process. That is, first the man takes a tiny step of length l_1, then the lion moves a tiny step, then the man takes a tiny step of length l_2, and so on. The man is free to do what he wants, but the lion is constrained to follow the "radius rule." We'll call the man's initial position M_0, and all his subsequent positions $M_1, M_2, M_3, M_4, \ldots$. In the same way, $L_0, L_1, L_2, L_3, \ldots$ will denote the lion's positions. Then, to quote Littlewood once more (the last sentence may well seem obscure, but I'll elaborate on it in just a bit),

> Starting from M's position at $t = 0$ there is a polygonal path $M_0, M_1, M_2 \ldots$ with the properties (i) $M_n M_{n+1}$ is perpendicular to OM_n, (ii) the total length is infinite, (iii) the path stays inside a circle round O inside the arena. In fact, if $l_n = M_{n-1} M_n$ [so $n = 1, 2, \ldots$] we have $OM_n^2 = OM_0^2 + \sum_{m=1}^{n} l_m^2$ and all is secure if we take $l_n = c n^{-3/4}$ with a suitable c.

Besicovitch's path is illustrated in figure 4.7.1, along with the lion's response at each step, under the assumption that the lion follows the "radius rule."

To understand figure 4.7.1, let me continue with Littlewood's next comments on Rado's problem:

> Let M run along this path (L keeping, as agreed, on OM). Since $M_0 M_1$ is perpendicular to $L_0 M_0$ [see figure 4.7.1], L does not catch M while M is on $M_0 M_1$. Since L_1 is on OM_1 [to be faithful to the "radius rule" L must do a tiny "side-jog" to the right, to L_1, as indicated by an arrowhead in figure 4.7.1] [and since] $M_1 M_2$ is perpendicular to $L_1 M_1$ [then] L does not catch M while M is on $M_1 M_2$. This continues for each successive $M_n M_{n+1}$, and for an infinite time since the total length [of the man's ploygonal path] is infinite.

There are several points in Littlewood's remarks that (in my opinion) are awfully terse, and that really demand some elaboration. One mathematician who discussed Littlewood's presentation (Otomar Hájek, in his 1975 book *Pursuit Games*) wrote that "Little can be added — the exposition is eminently clear," but I respectfully disagree! Let me then address these points, in what I think their order of increasing difficulty.

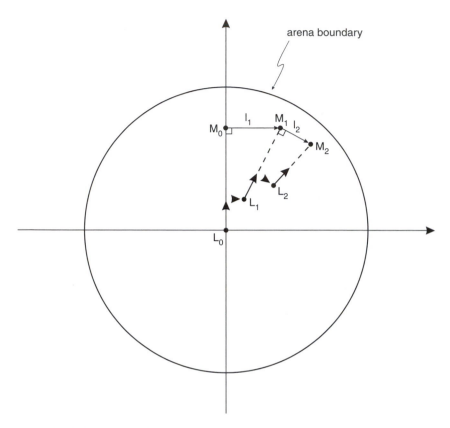

Figure 4.7.1 Besicovitch's path

(1) What does "a suitable c" mean? I'll return to this in (4), but for now it is sufficient simply to think of c as *extremely small*, making even the first step of the man ($l_1 = c$) extremely small; subsequent steps are even smaller. This makes plausible the claim that our turn-based, *discrete* stepping process of a lion chasing a man closely approximates a smooth, continuous process. If you need a specific number for c, then for now take $c = 10^{-(1000^{1000})}$. That's small!

(2) *Why* can't L catch M while M is on $M_n M_{n+1}$? The reason is pretty straightforward — L is running along OM_n, that is, toward where M *was* (remember, M moved first), *not* to where M *is* going.

(3) To see why the total length of M's path is infinite, observe (from figure 4.7.1) that we can write that length as

$$\sum_{n=1}^{\infty} l_n = \sum_{n=1}^{\infty} cn^{-3/4} = c \sum_{n=1}^{\infty} \frac{1}{n^{3/4}}.$$

This sum is known to diverge.[9] Thus, M's path is infinitely long for *any* $c > 0$, no matter how small.

(4) How do we know M's path always stays inside the arena? Applying the Pythagorean theorem repetitively to figure 4.7.1, we can write the following equations:

$$OM_0^2 + l_1^2 = OM_1^2$$
$$OM_1^2 + l_2^2 = OM_2^2$$
$$\vdots$$
$$OM_{n-2}^2 + l_{n-1}^2 = OM_{n-1}^2$$
$$OM_{n-1}^2 + l_n^2 = OM_n^2.$$

Then, successively substituting one equation into the next, starting with the *last* one, we have

$$OM_n^2 = l_n^2 + OM_{n-1}^2 = l_n^2 + l_{n-1}^2 + OM_{n-2}^2 = l_n^2 + l_{n-1}^2$$
$$+ \ldots + l_1^2 + OM_0^2$$

or, as Littlewood wrote,

$$OM_n^2 = OM_0^2 + \sum_{m=1}^{n} l_m^2.$$

Thus,

$$OM_n^2 = (\frac{1}{2})^2 + \sum_{m=1}^{n} (cm^{-3/4})^2 = \frac{1}{4} + c^2 \sum_{m=1}^{n} \frac{1}{m^{3/2}}.$$

That is, the distance of M from the center of the arena, OM, after n steps, is

$$OM_n = \sqrt{\frac{1}{4} + c^2 \sum_{m=1}^{n} \frac{1}{m^{3/2}}}.$$

Obviously, OM_n monotonically increases with n, and so the man spirals outward toward the arena wall as he runs. In the limit $n \to \infty$ the sum is known to be finite. So, as long as $\lim_{n \to \infty} OM_n \leq 1$ the man will always be inside the arena. This condition imposes an upper-bound on the value of c.

To end this chapter — and the book — here's one last challenge problem for you. Calculate a "reasonably good" upper bound on c (I'll leave it as part of the problem for you to ponder what "reasonably good" means). See appendix J for my answer.

Solution to the Challenge
Problems of Section 1.1

Our first question is easy to answer. To calculate the distance sailed by the pirate ship until it captures the merchant vessel ($n < 1$), simply recall from (1.1.12) that capture occurs at $(x_o, n/(1-n^2)x_o)$, i.e., the *merchant* has traveled a distance of $n/(1-n^2)x_o$. Since the pirate ship travels $1/n$ times faster than does the merchant, the pirate travels $1/n$ times as far, that is, the pirate ship travels a total distance of $1/(1-n^2)x_o$.

To answer the second question, i.e., to determine the distance the pirate ship lags behind the merchant vessel after a long time has passed (for $n = 1$), refer to figure A. There we see the pirate ship at point (x,y), while the merchant vessel is at (x_o, y_m). Note, carefully, that this is for any arbitrary time t. The distance separating the pirate ship and the merchant vessel is D, where

$$D^2 = (y_m - y)^2 + (x_o - x)^2 = (x_o - x)^2 \left[1 + \left(\frac{y_m - y}{x_o - x} \right)^2 \right].$$

Now, here's the crucial observation: the line joining (x,y) to (x_o, y_m) is the *tangent* to the pirate's pursuit curve, precisely because the chase

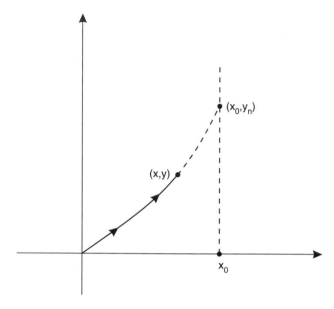

Figure A The geometry of the tail chase for n = 1

is a *pure* pursuit. Thus,

$$\frac{dy}{dx} = \frac{y_m - y}{x_o - x}$$

and so

$$D^2 = (x_o - x)^2 \left[1 + \left(\frac{dy}{dx} \right)^2 \right].$$

Substituting (1.1.9) for dy/dx for the $n = 1$ case, that is, writing

$$\frac{dy}{dx} = \frac{1}{2} \left[\frac{1}{(1 - x/x_o)} - \left(1 - \frac{x}{x_o} \right) \right],$$

we have

$$D^2 = (x_o - x)^2 \left[1 + \frac{1}{4} \left\{ \frac{1}{(1 - x/x_o)} - \left(1 - \frac{x}{x_o} \right) \right\}^2 \right]$$

$$= x_o^2 \left(1 - \frac{x}{x_o} \right)^2 \left[1 + \frac{1}{4} \left\{ \frac{1}{(1 - x/x_o)^2} - 2 + \left(1 - \frac{x}{x_o} \right)^2 \right\} \right]$$

$$= x_o^2 \left[\left(1 - \frac{x}{x_o} \right)^2 + \frac{1}{4} - \frac{1}{2} \left(1 - \frac{x}{x_o} \right)^2 + \frac{1}{4} \left(1 - \frac{x}{x_o} \right)^4 \right].$$

As $t \to \infty$ we physically see the pirate ship pull into behind the merchant and the pursuit become a vertically upward tail chase; thus $x \to x_o$ and so

$$\lim_{t \to \infty} D^2 = \lim_{x \to x_o} D^2 = \frac{1}{4} x_o^2$$

or, at last,

$$\lim_{t \to \infty} D = \frac{1}{2} x_o.$$

In Guha and Biswas (1994) you'll find an ingenious derivation of this result that does *not* require knowledge of (1.1.9), that is, of dy/dx. In their words, you can find $\lim_{t \to \infty} D$ *without* going through "a horrifying amount of calculus and algebra." But that "horrifying" stuff is of course the derivation of the pursuit curve itself, which strikes me as the *whole point* of the problem. It is precisely what you *don't* want to avoid!

Solutions to the Challenge
Problems of Section 1.2

For the total flight time T of the wind-blown plane, recall (1.2.1) and (1.2.8), where we showed that

$$\frac{dx}{dt} = -\frac{vx}{\sqrt{x^2 + y^2}}$$

and

$$y = \frac{a}{2}\left[\left(\frac{x}{a}\right)^{-n+1} - \left(\frac{x}{a}\right)^{n+1}\right], \quad n = \frac{w}{v}.$$

So,

$$\int_0^T dt = -\int_a^0 \frac{\sqrt{x^2 + y^2}}{vx} dx,$$

or

$$T = \frac{1}{v}\int_0^a \sqrt{1 + \frac{y^2}{x^2}} dx.$$

Also,

$$
\begin{aligned}
y^2 &= \frac{a^2}{4}\left[\left(\frac{x}{a}\right)^{-2n+2} - 2\left(\frac{x}{a}\right)^2 + \left(\frac{x}{a}\right)^{2n+2}\right] \\
&= \frac{a^2}{4}\left[\left(\frac{x}{a}\right)^{-2n}\frac{x^2}{a^2} - 2\frac{x^2}{a^2} + \left(\frac{x}{a}\right)^{2n}\frac{x^2}{a^2}\right].
\end{aligned}
$$

Thus,

$$
\frac{y^2}{x^2} = \frac{1}{4}\left[\left(\frac{x}{a}\right)^{-2n} - 2 + \left(\frac{x}{a}\right)^{2n}\right],
$$

and so

$$
\begin{aligned}
1 + \frac{y^2}{x^2} &= 1 + \frac{1}{4}\left[\left(\frac{x}{a}\right)^{-2n} - 2 + \left(\frac{x}{a}\right)^{2n}\right] \\
&= \frac{(x/a)^{-2n} - 2 + (x/a)^{2n} + 4}{4} \\
&= \frac{(x/a)^{-2n} + 2 + (x/a)^{2n}}{4} = \left\{\frac{(x/a)^n + (x/a)^{-n}}{2}\right\}^2.
\end{aligned}
$$

We can then write T as

$$
\begin{aligned}
T &= \frac{1}{2v}\int_0^a\left[\left(\frac{x}{a}\right)^n + \left(\frac{x}{a}\right)^{-n}\right]dx \\
&= \frac{1}{2v}\left[\int_0^a\left(\frac{x}{a}\right)^n dx + \int_0^a\left(\frac{x}{a}\right)^{-n} dx\right].
\end{aligned}
$$

Letting $u = x/a$ ($dx = a\,du$), we then have

$$
\begin{aligned}
T &= \frac{1}{2v}\left[\int_0^1 u^n a\,du + \int_0^1 u^{-n} a\,du\right] = \frac{a}{2v}\left[\frac{u^{n+1}}{n+1} + \frac{u^{-n+1}}{-n+1}\right]\Bigg|_0^1 \\
&= \frac{a}{2v}\left(\frac{1}{1+n} + \frac{1}{1-n}\right) = \frac{a/v}{1-n^2}, \quad n = \frac{w}{v}.
\end{aligned}
$$

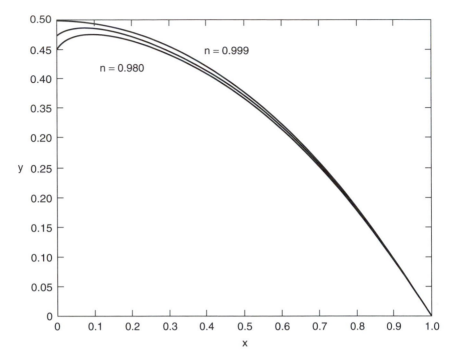

Figure B More paths of the wind-blown plane

This makes sense for $0 \leq n < 1$. Notice that if $n = 0$ (no wind) then $T = a/v$, which is simply the time the plane requires to fly straight along the x-axis from $(a,0)$ to $(0,0)$ at a speed v. As n approaches one from below, of course, we see $T \to \infty$ as expected.

For the total distance flown by the plane when n is "just less" than one, that is, for the case where the plane "just manages" to reach city C, recall that *at* $n = 1$ the plane's path is the parabola

$$ y = \frac{a}{2}\left[1 - \left(\frac{x}{a}\right)^2\right]. $$

As n approaches one, then, the upward curved part of the flight path of the plane approaches this parabola, as illustrated in figure B, which is an elaboration on figure 1.2.2. The three plots are for $n = 0.98$, $n = 0.99$ (the middle curve), and $n = 0.999$, all for $a = 1$.

From these curves it should be clear that the length of the longest flight path that *just* manages to reach city C is bounded from above by

$$\frac{a}{2} + \int_0^a \sqrt{1 + \left(\frac{dy}{dx}\right)^2}\, dx,$$

where the second term is the length of the parabolic arc. The first term, of course, is the length of the final leg of the journey back down along (or *almost* along, I should say) the vertical axis to city C at the origin. On the parabolic arc we have

$$\frac{dy}{dx} = -\frac{a}{2} 2 \left(\frac{x}{a}\right) \frac{1}{a} = -\frac{x}{a},$$

and so our answer is

$$\frac{a}{2} + \int_0^a \sqrt{1 + \left(\frac{x}{a}\right)^2}\, dx.$$

If we change variables to $u = x/a$ ($dx = a\, du$), our answer becomes

$$\frac{a}{2} + \int_0^1 \sqrt{1 + u^2 a}\, du = a \left[\frac{1}{2} + \int_0^1 \sqrt{1 + u^2}\, du\right]$$

or, with the aid of an integral table,

$$a \left[\frac{1}{2} + \left\{\frac{u\sqrt{u^2 + 1}}{2} + \frac{1}{2}\ln(u + \sqrt{u^2 + 1})\right\}\right]\Bigg|_0^1$$

$$= a \left[\frac{1 + \sqrt{2} + \ln(1 + \sqrt{2})}{2}\right] = 1.6478a.$$

This is the total distance flown by the plane when $n = 1 - \varepsilon$, where $\varepsilon > 0$ but arbitrarily small.

Our third challenge problem is essentially that of the wind-blown plane, except that we now have a *water*-blown dog! In the notation of figure 1.2.1, the wind speed w corresponds to the water speed (how nice for the notation that *wind* and *water* start with the same letter). Since we are told that the dog's still-water speed is 2 miles per hour (which corresponds to the plane's still-air speed v), the parameter $n = w/v$ is $w/2$. The equation of the dog's path is, again, from (1.2.8),

$$y(x) = \frac{a}{2}\left[\left(\frac{x}{a}\right)^{-n+1} - \left(\frac{x}{a}\right)^{n+1}\right],$$

where a now denotes the width of the stream. Since the dog starts at $x = a$, then it is two-thirds across the stream at $x = a/3$, at which point it is at its maximum downstream drift-point. That is,

$$\left.\frac{dy}{dx}\right|_{x=a/3} = 0.$$

Now,

$$\frac{dy}{dx} = \frac{a}{2}\left[(-n+1)\left(\frac{x}{a}\right)^{-n}\frac{1}{a} - (n+1)\left(\frac{x}{a}\right)^{n}\frac{1}{a}\right],$$

and so, when $x = a/3$ (and using $n = w/2$),

$$\left(-\frac{w}{2}+1\right)\left(\frac{1}{3}\right)^{-w/2} - \left(\frac{w}{2}+1\right)\left(\frac{1}{3}\right)^{w/2} = 0.$$

Or,

$$\frac{-w/2+1}{(\frac{1}{3})^{w/2}} = \left(\frac{w}{2}+1\right)\left(\frac{1}{3}\right)^{w/2}.$$

Or,

$$\frac{-w/2+1}{w/2+1} = \left(\frac{1}{3}\right)^{w} = \frac{1}{3^{w}} = \frac{2-w}{2+w}.$$

The solution is, *by inspection*, $w = 1$ (mile per hour).

To find the value of a, recall from (1.2.1) that the dog's speed in the x-direction is

$$\frac{dx}{dt} = -\frac{vx}{\sqrt{x^2+y^2}} = -\frac{2x}{\sqrt{x^2+y^2}}.$$

With $w = 1$ (so $n = \frac{1}{2}$) we have

$$y(x) = \left[\left(\frac{x}{a}\right)^{1/2} - \left(\frac{x}{a}\right)^{3/2}\right],$$

and so

$$x^2 + y^2 = x^2 + \frac{a^2}{4}\left[\frac{x}{a} - 2\frac{x^2}{a^2} + \frac{x^3}{a^3}\right]$$

$$= \frac{a^2}{4}\left[\frac{x}{a} + 2\frac{x^2}{a^2} + \frac{x^3}{a^3}\right] = \frac{a^2}{4}\left[\left(\frac{x}{a}\right)^{1/2} + \left(\frac{x}{a}\right)^{3/2}\right]^2.$$

Thus,

$$\frac{dx}{dt} = -\frac{2x}{(a/2)\left[(x/a)^{1/2} + (x/a)^{3/2}\right]},$$

Or

$$\frac{dt}{dx} = -\frac{a}{4}\left[\frac{(x/a)^{1/2} + (x/a)^{3/2}}{x}\right] = -\frac{1}{4}\left[\frac{(x/a)^{1/2} + (x/a)^{3/2}}{(x/a)}\right]$$

$$= -\frac{1}{4}\left[\frac{1}{(x/a)^{1/2}} + \left(\frac{x}{a}\right)^{1/2}\right].$$

If we call the time required for the dog to swim across the stream T, then integration gives us

$$4\int_0^T dt = 4T = -\int_a^0 \left[\frac{1}{(x/a)^{1/2}} + \left(\frac{x}{a}\right)^{1/2}\right]dx.$$

Changing variables to $u = x/a$ $(dx = a\,du)$ and noticing that $-\int\limits_{a}^{0} = \int\limits_{0}^{a}$, we have

$$4T = \int\limits_{0}^{1} \left[\frac{1}{u^{1/2}} + u^{1/2} \right] a\,du = a \left[2u^{1/2} + \frac{2}{3}u^{3/2} \right]_{0}^{1} = \frac{8}{3}a,$$

or

$$T = \frac{2}{3}a.$$

Now, the time it would take the dog to swim distance a *in still water*, at 2 miles per hour, is $T_o = a/2$, and since we are told that this is five minutes ($\frac{1}{12}$ hour) less than it actually takes in the flowing stream, we have

$$T - T_o = \frac{1}{12}.$$

So,

$$\frac{2}{3}a - \frac{1}{2}a = \frac{1}{12}.$$

I'll leave it for you to do the easy math that concludes from this that the stream is $a = \frac{1}{2}$ mile wide.

This sort of calculation may seem esoteric to many nonmathematical readers, but the *ideas* behind it have today percolated into the ken of even the broadly educated adult. As an example, consider the following passage from an essay by Ronald Reagan's biographer, written just after the former president's death (Edmund Morris, "The Unknowable: Ronald Reagan's Amazing, Mysterious Life," *The New Yorker*, June 28, 2004):

> In one of my last interviews with him, I [asked] Mr. President, do you realize that you had Einstein all figured out at age eighteen? There you were, a summer lifeguard [Reagan worked as a lifeguard during his college years] swaying every day in your high chair on the diving raft. Somebody starts to drown in midstream. You throw down your glasses — everything's a blur — you dive into the moving water — you swim, not to where the

drowning person is, but where he'll be by the time you intersect his trajectory. You think you're moving in a straight line. But actually you're describing a parabola, because the river's got you too. Your curve becomes his curve; you grab him, swing him around, and start back *in reverse*, not toward the diving platform but *upstream*, so that by the time you get to shallow water you'll be back where you started. During all this action, you're in a state of flux: no fixed point of reference, no sense of gravity. Everything's *relative* ...

Was Morris correct in saying the young Reagan's path was a parabola?

Solution to the Challenge
Problem of Section 1.5

To set up our analysis for this question, consider figure C1, which shows the geometry of the problem for the case $0 < \theta < 45°$ (which means, of course, $2\theta < 90°$). I treat the case of $\theta > 45°$ separately. If we call the altitude of the attacking missile h and its ground distance from the defensive site d (at the instant of the anti-missile's launch), then

$$\frac{h}{d} = \tan(\theta).$$

If we let T denote the time interval from launch until the supposed interception, then the attacking missile flys distance VT toward the defensive site and so, at the instant of supposed interception, the attacking missile is at a ground distance of $d - VT$. Thus,

$$\frac{h}{d - VT} = \tan(2\theta).$$

Finally, since the defensive anti-missile missile is launched at angle 2θ, its vertical speed component is $V \sin(2\theta)$; since the defensive missile must be at altitude h when interception occurs we must also have

$$h = VT \sin(2\theta).$$

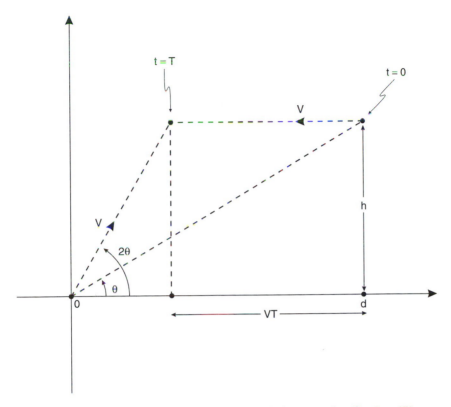

Figure C1 The geometry of missile, anti-missile interception for $\theta < 45°$

What our problem reduces to, then, is simply to show that the above three relations are not in conflict and, indeed, are consistent with each other no matter what the values of V, d, and h may be. This easy to do. From the first and third expressions we have

$$d = \frac{h}{\tan(\theta)} = \frac{VT\sin(2\theta)}{\tan(\theta)},$$

and so the second expression becomes

$$\frac{VT\sin(2\theta)}{VT\sin(2\theta)/\tan(\theta) - VT} = \tan(2\theta),$$

or, cancelling the VT factors on the left, we arrive at

$$\tan(2\theta) = \frac{\sin(2\theta)}{\sin(2\theta)/\tan(\theta) - 1}$$

which becomes, upon using the double-angle formula $\sin(2\theta) = 2 \sin(\theta) \cos(\theta)$, the assertion that

$$\tan(2\theta) = \frac{2 \sin(\theta) \cos(\theta)}{2 \sin(\theta) \cos(\theta)/\tan(\theta) - 1} = \frac{2 \tan(\theta)}{2 - \tan(\theta)/\sin(\theta) \cos(\theta)}$$

$$= \frac{2 \tan(\theta)}{2 - 1/\cos^2(\theta)}.$$

Is this true?

Yes, as you can see by writing

$$\frac{2 \tan(\theta)}{2 - 1/\cos^2(\theta)} = \frac{2 \tan(\theta)}{1 + 1 - 1/\cos^2(\theta)} = \frac{2 \tan(\theta)}{1 - (1 - \cos^2(\theta))/\cos^2(\theta)}$$

$$= \frac{2 \tan(\theta)}{1 - \sin^2(\theta)/\cos^2(\theta)} = \frac{2 \tan(\theta)}{1 - \tan^2(\theta)}$$

which is, indeed, equal to $\tan(2\theta)$, that is, it is the classic double-angle formula for $\tan(2\theta)$.

Now, what if $\theta > 45°$? The geometry of the problem is shown in figure C2, and from that figure we can write the expressions

$$\frac{h}{d} = \tan(\theta)$$

as before. Also, since the vertical speed component of the defensive missile is $V \sin(\pi - 2\theta)$, then, at the supposed interception at time $t = T$,

$$h = VT \sin(\pi - 2\theta) = VT[\sin(\pi) \cos(2\theta) - \cos(\pi) \sin(2\theta)],$$

or

$$h = VT \sin(\theta),$$

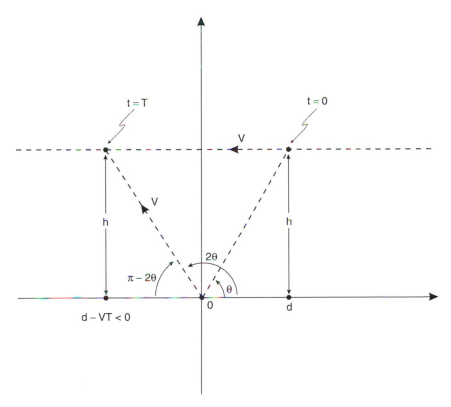

Figure C2 The geometry of missile, anti-missile interception for $\theta > 45°$

as before. And finally, from figure C2 we see that

$$\frac{h}{VT-d} = \tan(\pi - 2\theta) = \frac{\tan(\pi) - \tan(2\theta)}{1 + \tan(\pi)\tan(2\theta)} = -\tan(2\theta),$$

or

$$\frac{h}{d - VT} = \tan(2\theta),$$

as before. Thus, we have the same three expressions we had in the first analysis and so, of course, we arrive at the same conclusion.

Solution to the Challenge
Problem of Section 2.2

Referring back to the notation of figure 2.2.11, and reproducing (2.2.17), we have

$$\frac{dy}{dx} = \frac{x}{st - y}.$$

The length of the dog's run at time t is nst and so, if $y = y(x)$ is the equation of the dog's path, we have

$$nst = -\int_{1}^{x} \sqrt{1 + \left(\frac{dy}{dz}\right)^2}\, dz,$$

where the minus sign is explicitly included since the integral itself is *negative* (remember, we are interested in $0 \leq x \leq 1$), and yet nst must be positive. Solving each these two expressions for t, we have

$$t = -\frac{1}{ns}\int_{1}^{x} \sqrt{1 + \left(\frac{dy}{dz}\right)^2}\, dz = \frac{1}{s}\left[y + \frac{x}{dy/dx}\right],$$

or, with the definition

$$\frac{dy}{dx} = p(x),$$

this becomes

$$-\frac{1}{n} \int_1^x \sqrt{1 + p^2(z)}\, dz = y + \frac{x}{p(x)}.$$

Next, differentiating with respect to x,

$$-\frac{1}{n}\sqrt{1 + p^2(x)} = \frac{dy}{dx} + \frac{p(x) - x\, dp/dx}{p^2(x)} = p(x) + \frac{p(x) - x\, dp/dx}{p^2(x)}.$$

Doing some simple algebra soon gets us to the *separated* differential equation

$$\frac{dp}{p(1 + p^2) + (p^2/n)\sqrt{1 + p^2}} = \frac{dx}{x}.$$

The right-hand side is of course easy to integrate, but the left-hand side looks pretty scary! But looks are deceiving. If we make the change of variable

$$u = \frac{p}{\sqrt{1 + p^2}},$$

then it is easy to show that

$$dp = (1 + p^2)^{3/2}\, du.$$

Thus, the integration of the left-hand side of the "scary" equation becomes the far less scary

$$\int \frac{dp}{p(1 + p^2) + (p^2/n)\sqrt{1 + p^2}} = \int \frac{(1 + p^2)^{3/2} du}{p(1 + p^2) + (p^2/n)\sqrt{1 + p^2}}$$

$$= \int \frac{du}{\frac{p}{\sqrt{(1 + p^2)}} + (1/n) \cdot \frac{p^2}{1 + p^2}} = \int \frac{du}{u + (1/n)u^2}.$$

That is,

$$\int \frac{du}{u+(1/n)u^2} = \int \frac{dx}{x}, \quad u = \frac{p}{\sqrt{1+p^2}}.$$

The left-hand side in the last equation is easy to do if we make a partial fraction expansion:

$$\int \frac{du}{u+(1/n)u^2} = \int \frac{du}{u} - \frac{1}{n}\int \frac{du}{1+u/n}$$

and so, with C some arbitrary constant, we have

$$\ln(u) - \ln(1+u/n) + \ln(C) = \ln(x) = \ln\left(C\frac{u}{1+u/n}\right),$$

or

$$C\frac{u}{1+u/n} = x.$$

To determine the value of C, we can argue as follows. Looking at figure 2.2.11, it is geometrically obvious that $\lim_{x\to1-} p(x) = \lim_{x\to1-} \frac{dy}{dx} = -\infty$, where $x \to 1-$ means x approaches one from *below* (through values less than one). Thus, as $x \to 1-$ we see from the definition of u that $\lim_{x\to1-} u \to -1$, and so

$$C\frac{-1}{1-1/n} = 1,$$

or

$$C = \frac{1-n}{n},$$

and so

$$\frac{1-n}{n} \cdot \frac{u}{1+u/n} = x.$$

This is easily manipulated into the form $u(1 - n - x) = nx$ or, replacing u with its definition, we have

$$\frac{p}{\sqrt{1 + p^2}}(1 - n - x) = nx.$$

This, in turn, is easily manipulated into the form

$$p^2 = \frac{n^2 x^2}{(1 - n - x)^2 - n^2 x^2}.$$

And finally, taking the square root of both sides, and explicitly including a minus sign on the right because we know, *geometrically*, from figure 2.2.11, that $p = dy/dx \leq 0$ over the interval $0 \leq x \leq 1$, we can write

$$\frac{dy}{dx} = -\frac{nx}{\sqrt{(1 - n - x)^2 - n^2 x^2}}$$

$$= -\frac{nx}{\sqrt{(1 - n^2)x^2 - 2(1 - n)x + (1 - n)^2}}, \quad n > 1.$$

At this point all we need to finish our solution is a good table of integrals. After a little browsing in such a table you'll find that

$$\int \frac{x\,dx}{\sqrt{ax^2 + bx + c}} = \frac{\sqrt{ax^2 + bx + c}}{a} - \frac{b}{2a} \int \frac{dx}{\sqrt{ax^2 + bx + c}},$$

and

$$\int \frac{dx}{\sqrt{ax^2 + bx + c}} = -\frac{1}{\sqrt{-a}} \sin^{-1}\left(\frac{2ax + b}{\sqrt{b^2 - 4ac}}\right).$$

For our problem we have $a = 1 - n^2 < 0$, $b = -2(1 - n) > 0$, and $c = (1 - n)^2 > 0$, and so, with C once again some arbitrary constant,

$$y(x) = -n\frac{\sqrt{(1 - n^2)x^2 - 2(1 - n)x + (1 - n)^2}}{1 - n^2}$$

$$+ \frac{2(1 - n)n}{2(1 - n^2)} \cdot \frac{1}{\sqrt{n^2 - 1}} \sin^{-1}\left(\frac{2(1 - n^2)x - 2(1 - n)}{\sqrt{4(1 - n)^2 - 4(1 - n^2)(1 - n)^2}}\right) + C.$$

Since $y(x = 1) = 0$, it follows (with just a little algebra) that

$$C = \frac{n\pi}{2(n+1)\sqrt{n^2 - 1}}.$$

Then, setting $x = 0$ in the above expression for $y(x)$ we at last have our theoretical answer for the ordinate of the dog when it runs from right to left directly across the man's path:

$$y(0) = \frac{n}{n+1} + \frac{n}{(n+1)\sqrt{n^2 - 1}}\sin^{-1}(1/n) + \frac{n\pi}{2(n+1)\sqrt{n^2 - 1}}.$$

Notice that $\lim_{n\to\infty} y(0) = 1$, which makes intuitive sense; an *infinitely* fast dog will run around the man along a *circle* with radius one, and so clearly $y(0) = 1$. The fact that the boxed expression has this property does not, of course, prove that the expression is correct in general for all $n > 1$, but if it did *not* have this property that *would* prove the expression to be *incorrect*. So, it is a useful check to make.

Now, with the above answer written down, let me admit that at various points in the last several steps there are a number of sign ambiguities due to the square-root operations, that is, at each step do we use the positive root or the negative root? If one goes through the above calculations with a specific value for n, rather than with a literal n, it is a bit easier to get a "feel" for which root to use at each step, but this claim is not likely to convince you that the boxed expression for $y(0)$ is the result of correct decisions with no slip-ups. So, here's another way to convince you (and me!) of the correctness of our answer. Morley's problem is a well-defined *physical process*, so let's simply *physically simulate* the dog's running path on a computer and *watch* where the path, for various values of n, crosses the y-axis. Then, if the theoretical $y(0)$ expression in the box is correct, it should give pretty nearly the same numerical values as does the simulation (almost certainly there will be *some* discrepancy because a computer simulation will invariably suffer from cumulative round-off errors).

To see how to do a computer simulation of the dog's running path, consider the simple program flow-chart of figure D1 (with reference to

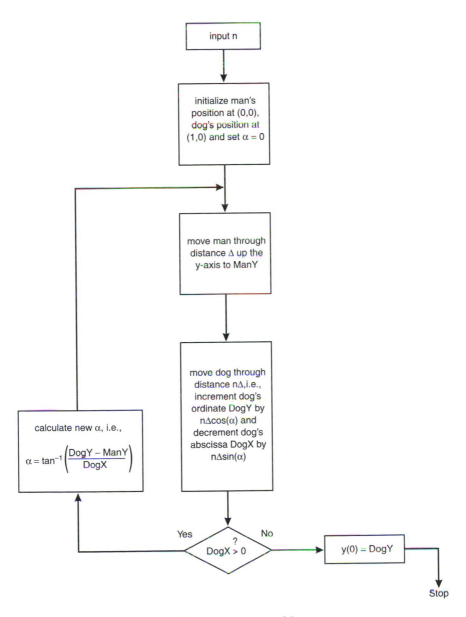

Figure D1 The logic of simulating Morley's problem

figure D2, α is the angle the line connecting D and M makes with the horizontal and so, as the dog's motion is always perpendicular to this line, the dog's motion also makes angle α with the vertical).

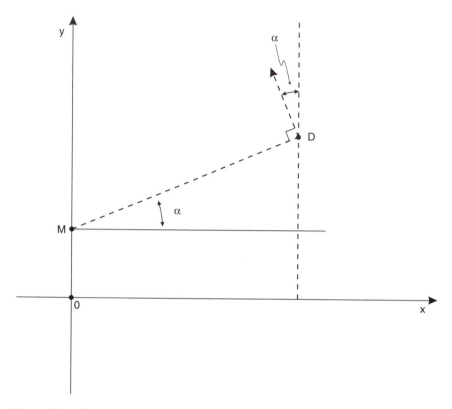

Figure D2 The geometry of simulating Morley's problem

I implemented the logic of figure D1 (which is *Euler's method* — see note 2 for chapter 4), as well as the formula for $y(0)$, each in MATLAB code, and ran each program for five different values of n (in the simulation code I used a step-increment of $\Delta = 10^{-4}$). The results were

	$y(0)$	
n	simulation	theoretical
1.5	1.8347	1.8346
2	1.4730	1.4728
2.5	1.3325	1.3322
3	1.2569	1.2566
10	1.0626	1.0618

It's a subjective evaluation, of course, but *I'd* say this is pretty good agreement between theory and "experiment."

Solution to the Challenge
Problem of Section 2.3

From the analysis done in section 2.3 we have, for $k = 2$, the following results: from (2.3.8):

$$r = R_0 \frac{\cos(\theta/2)}{\cos(\theta_0/2)} \cdot \frac{\sin^3(\theta_0/2)}{\sin^3(\theta/2)}$$

and, from immediately after (2.3.10),

$$\lim_{\theta \to \pi, \theta_0 \neq 0} \frac{d\theta}{dt} = 2 \frac{V_t}{R_0} \cdot \frac{\cos(\theta_0/2)}{\sin^3(\theta_0/2)}.$$

From the same arguments given in section 2.3, which made no use of any specific value for k, we also have *in the terminal phase* of the attack that

$$\frac{d^2 r}{dt^2} = 0$$

and, for $k = 2$, that

$$\frac{dr}{dt} = -V_t.$$

Thus, the missile's terminal *radial* acceleration is

$$\left[\frac{d^2r}{dt^2} - r\left(\frac{d\theta}{dt}\right)^2\right]_{\theta=\pi} = -R_0 \frac{\cos(\pi/2)}{\cos(\theta_0/2)} \cdot \frac{\sin^3(\theta_0/2)}{\sin^3(\pi/2)} \cdot 4\frac{V_t^2}{R_0^2} \cdot \frac{\cos^2(\theta_0/2)}{\sin^6(\theta_0/2)} = 0$$

since $\cos(\pi/2) = 0$.

For the *transverse* acceleration, we need to know $d^2\theta/dt^2$. Looking back at (2.3.10), for $k = 2$, we have for *arbitrary* θ that

$$\frac{d\theta}{dt} = 2\frac{V_t}{R_0} \cdot \frac{\cos(\theta_0/2)}{\sin^3(\theta_0/2)} \cdot \sin^4(\theta/2)$$

which I've put in a box because I'm going to refer back to this expression in just a bit. From this we can immediately write

$$\frac{d^2\theta}{dt^2} = 2\frac{V_t}{R_0} \cdot \frac{\cos(\theta_0/2)}{\sin^3(\theta_0/2)} \cdot 4\sin^3(\theta/2)\cos(\theta/2) \cdot \frac{1}{2}\frac{d\theta}{dt}$$

or, replacing the $d\theta/dt$ factor with its equivalent in the above box,

$$\frac{d^2\theta}{dt^2} = 8\frac{V_t^2}{R_0^2} \cdot \frac{\cos^2(\theta_0/2)}{\sin^6(\theta_0/2)} \sin^7(\theta/2)\cos(\theta/2)$$

which is zero in the terminal phase ($\theta = \pi$) of the attack. So, the missile's terminal *transverse* acceleration is

$$2\frac{dr}{dt} \cdot \frac{d\theta}{dt} + r\frac{d^2\theta}{dt^2} = 2(-V_t)2\frac{V_t}{R_0} \cdot \frac{\cos(\theta_0/2)}{\sin^3(\theta_0/2)},$$

which means the answer to our first challenge calculation is

$$\text{terminal transverse acceleration} = -\frac{4V_t^2}{R_0} \cdot \frac{\cos(\theta_0/2)}{\sin^3(\theta_0/2)}, \quad 0 < \theta_0 < \pi.$$

Ignoring the minus sign (which simply means that the acceleration vector is in the direction *opposite* to the direction of the position vector, that is, is pointed *toward* the target, which makes physical sense), the terminal transverse acceleration for $R_0 = 25,000$ feet (about five miles), $V_t = 1,000$ feet per second (since $k = 2$ this means the missile speed is 2,000 feet per second), and a broadside launch angle of $\theta_0 = 90° = \pi/2$ radians is

$$\frac{4(1,000\,\text{ft/sec})^2}{25,000\,\text{ft}} \cdot \frac{\cos(\pi/4)}{\sin^3(\pi/4)} = \frac{4,000,000}{25,000} \cdot \frac{1/\sqrt{2}}{(1/\sqrt{2})^3}\,\text{ft/sec}^2 = 320\,\text{ft/sec}^2,$$

or, as one "gee" is $32\,\text{ft/sec}^2$, the terminal transverse acceleration is 10 gees. This is well within the capability of modern air-to-air intercept missiles, for example, the more than three decades-old Hughes/Raytheon Defense Systems AIM-9J Sidewinder can execute a 22-gee transverse acceleration maneuver at sea level (loss of aerodynamic control forces in the thin air at high altitudes reduces this capability to 13 gees at an altitude of 50,000 feet, but that is still enough for the 10-gee requirement to be satisfied).

Next, to calculate the total missile flight time from launch to impact, once again return to the boxed expression for $d\theta/dt$, which allows us to write

$$dt = \frac{R_0}{2V_t} \cdot \frac{\sin^3(\theta_0/2)}{\cos(\theta_0/2)} \cdot \frac{1}{\sin^4(\theta/2)}\,d\theta.$$

Or, writing $C = \dfrac{R_0}{2V_t} \cdot \dfrac{\sin^3(\theta_0/2)}{\cos(\theta_0/2)}$,

$$dt = \frac{C}{\sin^4(\theta/2)}\,d\theta.$$

If we call the total flight time T, then

$$\int_0^T dt = T = C \int_{\theta_0}^{\pi} \frac{1}{\sin^4(\theta/2)}\,d\theta.$$

From integral tables we have

$$\int \frac{dx}{\sin^n(ax)} = -\frac{\cos(ax)}{a(n-1)\sin^{n-1}(ax)} + \frac{n-2}{n-1}\int \frac{dx}{\sin^{n-2}(ax)}$$

and

$$\int \frac{dx}{\sin^2(ax)} = -\frac{1}{a}\cot(ax).$$

So,

$$\int_{\theta_0}^{\pi} \frac{1}{\sin^4(\theta/2)}d\theta = -\frac{\cos(\theta/2)}{\frac{1}{2}3\sin^3(\theta/2)}\bigg|_{\theta_0}^{\pi} + \frac{2}{3}\int_{\theta_0}^{\pi} \frac{d\theta}{\sin^2(\theta/2)}$$

$$= \frac{\cos(\theta_0/2)}{\frac{3}{2}\sin^3(\theta_0/2)} + \frac{2}{3}\left[-2\cot\left(\frac{\theta}{2}\right)\right]_{\theta_0}^{\pi} = \frac{2\cos(\theta_0/2)}{3\sin^3(\theta_0/2)}$$

$$-\frac{4}{3}\left[\cot\left(\frac{\pi}{2}\right) - \cot\left(\frac{\theta_0}{2}\right)\right] = \frac{2\cos(\theta_0/2)}{3\sin^3(\theta_0/2)} + \frac{4\cos(\theta_0/2)}{3\sin(\theta_0/2)}$$

$$= \frac{2\cos(\theta_0/2) + 4\sin^2(\theta_0/2)\cos(\theta_0/2)}{3\sin^3(\theta_0/2)} = \frac{2\cos(\theta_0/2)[1+2\sin^2(\theta_0/2)]}{3\sin^3(\theta_0/2)}.$$

Therefore,

$$T = \frac{R_0}{2V_t} \cdot \frac{\sin^3(\theta_0/2)}{\cos(\theta_0/2)} \cdot \frac{2\cos(\theta_0/2)[1+2\sin^2(\theta_0/2)]}{3\sin^3(\theta_0/2)}$$

or,

$$\boxed{T = \frac{R_0}{3V_t}\left[1+2\sin^2\left(\frac{1}{2}\theta_0\right)\right],}$$

a result that gives us the answers to the last two challenge calculations.

That is, for a broadside launch ($\theta_0 = \pi/2$) we have

$$T = \frac{2}{3} \cdot \frac{R_0}{V_t},$$

and for a launch angle of $\theta_0 = 60° = \pi/3$ we have

$$T = \frac{1}{2} \cdot \frac{R_0}{V_t}.$$

We see that the $\theta_0 = 60°$ launch angle has a significantly *smaller* flight time (for the same R_0 and V_t) compared to a broadside launch. This makes sense, too, I think — a missile launched broadside is *always chasing after* the target, while a missile launched at an angle of $\theta_0 = 60°$ has a significant portion of its flight with the target traveling *toward* the missile.

These are interesting and useful calculations to make since any real missile has available only a finite amount of rocket fuel for its engine, and electrical energy for the guidance electronics; for example, the AIM-9J Sidewinder has a guidance power duration of 40 seconds. For our numerical example, with $R_0 = 25,000$ feet and $V_t = 1,000$ feet/second, our two values of T are 17 seconds (for $\theta_0 = 90°$) and 12.5 seconds (for $\theta_0 = 60°$), so it *appears* that the Sidewinder could fly either attack profile. But that is not so, because the $\theta_0 = 60°$ launch angle has a subtle flaw — it requires a substantially greater terminal transverse acceleration compared to the broadside launch:

$$\frac{4V_t^2}{R_0} \cdot \frac{\cos(30°)}{\sin^3(30°)} = \frac{4,000,000 \, \text{ft}^2/\text{sec}^2}{25,000 \, \text{ft}} \cdot \frac{\sqrt{3}/2}{\left(\dfrac{1}{2}\right)^3}$$

$$= 1,109 \, \text{ft/sec}^2 = 36.4 \, \text{gees}.$$

Even in the dense air at sea level, with maximum aerodynamic forces available, the Sidewinder could not fly such a (pure pursuit) attack.

Solution to the Challenge
Problem of Section 2.5

From equation (2.5.8) we have the length r of the position vector to the coast-guard boat, on its outward spiral path, as

$$r = d\,\frac{e^{\theta/\lambda}}{1+(k_2/k_1)}, \quad \lambda = \sqrt{\left(\frac{k_2}{k_1}\right)^2 - 1},$$

where d is the initial distance between the rum runner and the coast-guard boat, and k_1 and k_2 are the speeds of the rum runner and the coast-guard boat, respectively. In polar coordinates the differential arc length ds satisfies $(ds)^2 = (dr)^2 + (rd\theta)^2$, and so the length of the spiral path travelled by the coast guard boat is

$$\int ds = \int \sqrt{(dr)^2 + (rd\theta)^2} = \int \sqrt{\left(\frac{dr}{d\theta}\right)^2 + r^2}\,d\theta.$$

Since

$$\frac{dr}{d\theta} = \frac{d}{\lambda} \cdot \frac{e^{\theta/\lambda}}{1+(k_2/k_1)} = \frac{r}{\lambda},$$

then our path-length integral is, under the worst-case scenario specified in the problem statement (the coast-guard boat has to make nearly one complete swing around the origin to make its interception of the rum runner),

$$
\int_0^{2\pi} \sqrt{\frac{r^2}{\lambda^2} + r^2} \, d\theta = \int_0^{2\pi} r \sqrt{\frac{1}{\lambda^2} + 1} \, d\theta = \sqrt{\frac{1}{\lambda^2} + 1} \int_0^{2\pi} d \frac{e^{\theta/\lambda}}{1 + (k_2/k_1)} \, d\theta
$$

$$
= \frac{d\sqrt{(1+\lambda^2)}/\lambda}{1 + (k_2/k_1)} (\lambda e^{\theta/\lambda}|_0^{2\pi} = \frac{d\sqrt{1+\lambda^2}}{1 + (k_2/k_1)} (e^{2\pi/\lambda} - 1).
$$

Since $\sqrt{1+\lambda^2} = \sqrt{1 + (k_2/k_1)^2 - 1} = k_2/k_1$, then the spiral portion of the coast-guard boat's path length is

$$
\frac{d \, k_2/k_1}{1 + (k_2/k_1)} (e^{2\pi/\lambda} - 1),
$$

and the time it takes the coast-guard boat to travel this distance is

$$
\frac{\frac{d(k_2/k_1)}{1+(k_2/k_1)} (e^{2\pi/\lambda} - 1)}{k_2} = \frac{d}{k_1 + k_2} (e^{2\pi/\lambda} - 1).
$$

You'll recall, from just before (2.5.1), that we wrote the time for the coast-guard boat to travel from its location where it initially spotted the rum runner, to where it starts its outward spiral path, as

$$
T = \frac{d}{k_1 + k_2},
$$

and so the time required by the coast-guard boat to make one complete swing around the origin is $T(e^{2\pi/\lambda} - 1)$. Thus, the *total* time, $\widehat{T}$, from initial sighting to interception, is

$$
\widehat{T} = T(e^{2\pi/\lambda} - 1) + T = Te^{2\pi/\lambda} = \frac{d}{k_1} \cdot \frac{e^{2\pi/\lambda}}{1 + (k_2/k_1)} = \frac{d}{k_1} \cdot \frac{e^{2\pi/\lambda}}{1 + 1.2},
$$

because we are told that $k_2/k_1 = 1.2$.

If we now increase the speed ratio (by making k_2 larger and keeping k_1 fixed) to $k_2/k_1 = 1.4$, we have the new total time from first sighting to interception — let's call it N — given by

$$N = \frac{d}{k_1} \cdot \frac{e^{2\pi/\lambda_N}}{1 + 1.4}, \quad \lambda_N = \sqrt{(1.4)^2 - 1}.$$

Notice that the factor d/k_1 is a *constant*, as both d and the rum runner's speed are unchanged. So,

$$\frac{\widehat{T}}{N} = \frac{e^{2\pi/\lambda}}{1 + 1.2} \cdot \frac{1 + 1.4}{e^{2\pi/\lambda_N}} = \frac{2.4}{2.2} \cdot \exp\left[2\pi\left(\frac{1}{\lambda} - \frac{1}{\lambda_N}\right)\right].$$

Since

$$\frac{1}{\lambda} - \frac{1}{\lambda_N} = \frac{1}{\sqrt{(1.2)^2 - 1}} - \frac{1}{\sqrt{(1.4)^2 - 1}} = \frac{1}{\sqrt{0.44}} - \frac{1}{\sqrt{0.96}} = 0.487,$$

then

$$\frac{\widehat{T}}{N} = \frac{2.4}{2.2} \cdot e^{2\pi(0.487)} = 23.26.$$

That is, the total time from first sighting to interception is reduced by a factor of 23.25 by increasing the coast-guard boat's speed advantage from 20% to 40%. Do you find this quite large reduction factor surprising?

Solution to the Challenge
Problem of Section 3.2

Looking back at (3.2.1),

$$\frac{dr}{dt} = -v \sin\left(\frac{\pi}{n}\right),$$

which, as $r(t = 0) = r_0$, integrates immediately to

$$r(t) = r_0 - vt \sin\left(\frac{\pi}{n}\right).$$

Also, from (3.2.2), and using our result for $r(t)$,

$$\frac{d\theta}{dt} = \frac{v \cos(\pi/n)}{r} = \frac{v \cos(\pi/n)}{r_0 - vt \sin(\pi/n)}.$$

Integrating indefinitely, we have

$$\theta(t) = v \cos\left(\frac{\pi}{n}\right) \int \frac{dt}{r_0 - vt \sin(\pi/n)} + C.$$

Changing variables to $u = r_0 - vt \sin(\pi/n)$, that is, $du = -v \sin(\pi/n) dt$, we have

$$\theta(t) = v \cos\left(\frac{\pi}{n}\right) \int \frac{-du}{uv \sin(\pi/n)} + C = -\cot\left(\frac{\pi}{n}\right) \int \frac{du}{u} + C$$

$$= -\cot\left(\frac{\pi}{n}\right) \ln(u) + C,$$

or

$$\theta(t) = -\cot\left(\frac{\pi}{n}\right) \ln\left[r_0 - vt \sin\left(\frac{\pi}{n}\right)\right] + C.$$

Since $\theta(t=0) = 0$, we have $C = \cot(\pi/n) \ln(r_0)$ or, at last,

$$\theta(t) = -\cot\left(\frac{\pi}{n}\right) \ln\left[\frac{r_0 - vt \sin(\pi/n)}{r_0}\right].$$

As derived in (3.2.5), the total duration of the n-bug cyclic pursuit is

$$T = \frac{r_0}{v \sin(\pi/n)},$$

and so if $\widehat{t} = 0.99T$ we have

$$\theta(\widehat{t}) = -\cot\left(\frac{\pi}{n}\right) \ln\left[\frac{r_0 - 0.99r_0}{r_0}\right] = -\frac{\ln(0.01)}{\tan\left(\frac{\pi}{n}\right)},$$

or

$$\theta(\widehat{t}) = \frac{\ln(100)}{\tan(\pi/n)}.$$

If $n = 1,000$ (giving a regular polygon virtually indistinguishable from a circle), then the number of swings around the center of the 1000-gon during the first 99% of the total pursuit time is

$$\frac{\theta(\widehat{t})}{2\pi} = \frac{\ln(100)}{2\pi \tan(\pi/1,000)} = 233.3.$$

Solution to the Challenge
Problem of Section 4.3

Using the hint provided by Professor Bailey, we write the differential arc length ds along the rabbit's path, in polar coordinates, as satisfying

$$\left(\frac{ds}{d\theta}\right)^2 = r^2 + \left(\frac{dr}{d\theta}\right)^2.$$

In this system of coordinates the initial value of θ is $\pi/2$, because you'll recall that the rabbit is on the y-axis at $(0, c)$ at time $t = 0$. Using the notation just as presented in section 4.3, we have (look back in the text to just before (4.3.1)) that

$$\frac{ds}{dt} = V_r.$$

The chain rule from calculus then tells us that

$$\frac{ds}{d\theta} = \frac{ds}{dt} \cdot \frac{dt}{d\theta} = \frac{ds/dt}{d\theta/dt} = \frac{V_r}{d\theta/dt}.$$

Now, again as before (look back in the text to just before (4.3.2)), we have

$$\tan\{\theta(t)\} = \frac{b}{V_f t},$$

and so, as

$$\frac{d}{dt}\tan\{\theta(t)\} = \frac{d}{d\theta}\tan(\theta) \cdot \frac{d\theta}{dt} = \frac{1}{\cos^2(\theta)} \cdot \frac{d\theta}{dt},$$

then

$$\frac{d\theta}{dt} = \cos^2(\theta)\frac{d}{dt}\tan(\theta) = \cos^2(\theta)\frac{d}{dt}\left(\frac{b}{V_f t}\right) = -\frac{b}{V_f t^2}\cos^2(\theta)$$

$$= -\frac{V_f}{b} \cdot \frac{b^2}{V_f^2 t^2}\cos^2(\theta) = -\frac{V_f}{b}\tan^2(\theta)\cos^2(\theta) = -\frac{V_f}{b}\sin^2(\theta).$$

Thus,

$$\frac{ds}{d\theta} = \frac{V_r}{-(V_f/b)\sin^2(\theta)} = -\frac{bV_r}{V_f} \cdot \frac{1}{\sin^2(\theta)}.$$

So, returning to our very first expression for the differential arc-length,

$$\left\{-\frac{bV_r}{V_f} \cdot \frac{1}{\sin^2(\theta)}\right\}^2 = r^2 + \left(\frac{dr}{d\theta}\right)^2,$$

that is,

$$\frac{b^2 V_r^2}{V_f^2} \cdot \frac{1}{\sin^4(\theta)} = r^2 + \left(\frac{dr}{d\theta}\right)^2.$$

Or, multiplying through by $V_f^2/b^2 V_r^2$, we have

$$\csc^4(\theta) = \left(\frac{rV_f}{bV_r}\right)^2 + \left[\frac{d}{d\theta}\left\{\frac{rV_f}{bV_r}\right\}\right]^2.$$

Then, following Professor Bailey's suggestion and defining the dimensionless quantity

$$R = \frac{r V_f}{b V_r},$$

we have

$$\csc^4(\theta) = R^2 + \left[\frac{d R}{d\theta}\right]^2,$$

so, at last, we arrive at Professor Bailey's differential equation

$$\frac{d R}{d\theta} = \pm\sqrt{\csc^4(\theta) - R^2},$$

and we are done.

The advantage of the polar coordinate formulation that we have just done is that we have only one dimensionless quantity, rather than the two we had for the rectangular coordinate formulation in chapter 4. We *have* paid a price for this reduction, however. The ease with which we determined the nature of the solutions in chapter 4 does not seem to carry over into the polar coordinate formulation. The lesson here? The ancient dictum that "there is no such thing as a free lunch" is as true in mathematics as it is in economics!

Appendix I

*Solution to the Challenge
Problem of Section 4.4*

If **B** and **R** use probability distributions $(p,\ 1-p)$ and $(q,\ 1-q)$, respectively, then from the pay-off matrix of figure 4.4.6 the average game value can be written as either

$$V(p,q) = p[4 \cdot q + 1 \cdot (1-q)] + (1-p)[3 \cdot q + 4 \cdot (1-q)],$$

or

$$V(p,q) = q[4 \cdot p + 3 \cdot (1-p)] + (1-q)[1 \cdot p + 4 \cdot (1-p)],$$

with both expressions reducing to

$$V(p,q) = 4 - 3p - q + 4pq.$$

Figure I clearly shows a minimax point, which can be determined from

$$\frac{\partial V}{\partial p} = 0 = -3 + 4q$$

and

$$\frac{\partial V}{\partial q} = 0 = -1 + 4p.$$

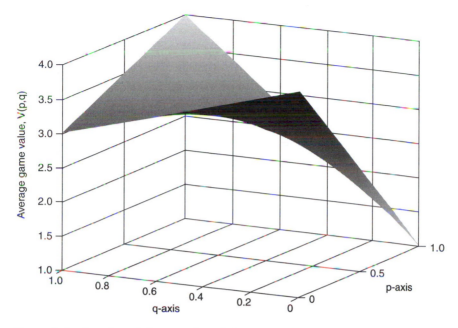

Figure I Seeing the minimax point of the attack-defense game

So, the optimal probability distribution for **B**'s mixed strategy is $(\frac{1}{4}, \frac{3}{4})$ and the optimal probability distribution for **R**'s mixed strategy is $(\frac{3}{4}, \frac{1}{4})$. That is, **B** should defend the more valuable installation with three times the probability that he defends the less valuable one — this is perhaps "intuitively obvious" — but **R** should attack the *less* valuable installation with three times the probability that he attacks the *more* valuable one, a conclusion that I think not at all obvious. The average value of the game at the minimax point is

$$V\left(\frac{1}{4}, \frac{3}{4}\right) = 4 - 3\left(\frac{1}{4}\right) - \left(\frac{3}{4}\right) + 4\left(\frac{1}{4}\right)\left(\frac{3}{4}\right)$$

$$= 4 - \left(\frac{3}{4}\right) - \left(\frac{3}{4}\right) + \left(\frac{3}{4}\right) = 3.25.$$

Appendix J

For the man to remain in the arena, we have

$$\sqrt{\frac{1}{4} + c^2 \sum_{m=1}^{\infty} \frac{1}{m^{3/2}}} \leq 1$$

and so the upper-bound on c is given by

$$c \leq \frac{1}{2}\sqrt{3 \Big/ \sum_{m=1}^{\infty} \frac{1}{m^{3/2}}} = \frac{1}{2}\sqrt{3} \Big/ \sqrt{\sum_{m=1}^{\infty} \frac{1}{m^{3/2}}}.$$

We will have to evaluate the sum numerically because there is no known exact result for it. We will have to use a bit of care in doing that as the partial sums converge very slowly. We do know, from a 1734 result due to Euler, that

$$\sum_{m=1}^{\infty} \frac{1}{m^2} = \frac{\pi^2}{6} = 1.644934\ldots,$$

and so our sum — term-by-term larger than those in Euler's sum — will be *larger*. But how much larger? A quick-and-dirty computer calculation "suggests" a result of about 2.6, but even using hundreds of thousands of terms shows how slow the convergence is (the first 1,000 terms sum to 2.5491, the first 10,000 terms sum to 2.5924, the first 100,000 terms sum to 2.6061, and the first 1,000,000 terms sum to 2.6173). Our caution on this matter is prompted by knowledge of how very slowly the harmonic series grows, and yet its partial sums eventually exceed *any* finite value. We know our sum doesn't do that (look at note 9 for chapter 4), but just how big *can* its partial sums get? Can they get significantly larger than 2.6? We need to be only just a bit clever to answer that (the answer is *no*), about as clever as is the typical graduate of first-year calculus!

Suppose $f(x)$ is any positive, continuous function that decreases as x increases, and has values $f(m) = a_m$ for integer values (m) for x. Then, with reference to figure J, and using the area interpretation of the integral, we can write

$$\int_1^\infty f(x)dx \geq a_2 + a_3 + a_4 + \ldots,$$

or

$$a_1 + \int_1^\infty f(x)dx \geq \sum_{m=1}^\infty a_m = \sum_{m=1}^\infty f(m).$$

Also,

$$\int_1^\infty f(x)dx \leq a_1 + a_2 + a_3 + a_4 + \cdots = \sum_{m=1}^\infty f(m).$$

So,

$$\int_1^\infty f(x)dx \leq \sum_{m=1}^\infty f(m) \leq f(1) + \int_1^\infty f(x)dx.$$

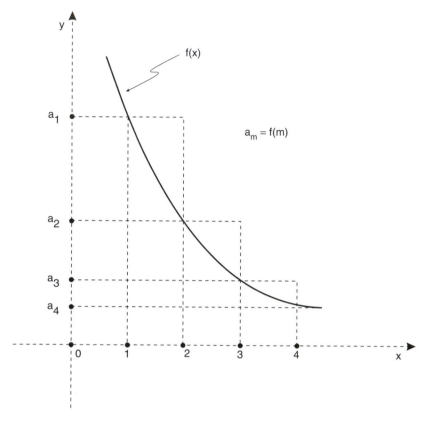

Figure J Bounding a sum with an integral

If $f(x) = 1/x^{3/2}$, for example, then since $f(1) = 1$ we have

$$\int_1^\infty \frac{dx}{x^{3/2}} \leq \sum_{m=1}^\infty \frac{1}{m^{3/2}} \leq 1 + \int_1^\infty \frac{dx}{x^{3/2}}.$$

And since

$$\int_1^\infty \frac{dx}{x^{3/2}} = -2x^{-1/2} \Big|_1^\infty = 2$$

we have

$$2 \le \sum_{m=1}^{\infty} \frac{1}{m^{3/2}} \le 3$$

which agrees with the computer "suggestion" that the sum is about 2.6.

If we use the upper bound of 3 on the sum we will arrive at a conservative estimate for the largest c may be (by conservative I mean that an even larger sum would keep the man inside the arena). So, using 3, we have

$$c \le \frac{\sqrt{3}/2}{\sqrt{3}} = \frac{1}{2} = 0.5,$$

We can get tighter bounds on the sum in the same way (and, hence, a larger, that is, a *less* restrictive, upper bound on c) by looking again at figure J. Thus,

$$\int_{2}^{\infty} f(x)dx \ge a_3 + a_4 + \cdots$$

or,

$$a_1 + a_2 + \int_{2}^{\infty} f(x)dx \ge \sum_{m=1}^{\infty} f(m).$$

Also,

$$\int_{2}^{\infty} f(x)dx \le a_2 + a_3 + \cdots,$$

or

$$a_1 + \int_{2}^{\infty} f(x)dx \le \sum_{m=1}^{\infty} f(m).$$

Thus,

$$f(1) + \int\limits_2^\infty f(x)dx \le \sum_{m=1}^\infty f(m) \le f(1) + f(2) + \int\limits_2^\infty f(x)dx.$$

So,

$$1 + \int\limits_2^\infty \frac{dx}{x^{3/2}} \le \sum_{m=1}^\infty f(m) \le 1 + \frac{1}{2\sqrt{2}} + \int\limits_2^\infty \frac{dx}{x^{3/2}}.$$

Since

$$\int\limits_2^\infty \frac{dx}{x^{3/2}} = -2x^{-1/2} \mid_2^\infty = \frac{2}{\sqrt{2}} = \sqrt{2},$$

then

$$1 + \sqrt{2} \le \sum_{m=1}^\infty \frac{1}{m^{3/2}} \le 1 + \frac{1}{2\sqrt{2}} + \sqrt{2},$$

or

$$2.414 \le \sum_{m=1}^\infty \frac{1}{m^{3/2}} \le 2.768.$$

Recomputing our upper bound on c, we have

$$c \le \frac{\sqrt{3/2}}{\sqrt{2.768}} = 0.52.$$

We could continue this process indefinitely by using figure J over and over to get ever tighter bounds on the sum. But since our increase on the upper bound for c was so little (just 4%) with the last calculation, another iteration would produce an even less impressive increase, and so I think what we already have is a "reasonably good" estimate for the smallest upper bound on c.

Guelman's Proof

The problem I'm going to discuss appeared in a 1971 paper — see Guelman (1971) — written by an electrical engineer who was studying a class of problems in missile guidance theory, but for us it will be a pure mathematics problem. The problem itself is easy to state: if k and ν are each positive constants such that $\nu > 1$ and $k\nu > 1$, and if ϕ_0 is an arbitrary constant, then show that the equations

(a) $\nu \cos(\phi_0 - k\theta) - \cos(\theta) = 0,$
(b) $\nu \sin(\phi_0 - k\theta) - \sin(\theta) = 0$

have what are called *intercalated roots*. That is, the infinity of solutions to each of (a) and (b) *alternate* (are *interwoven*) along the θ-axis. Formally, our problem is to prove that, if the solutions to (a) are $\ldots, \theta_{a_{j-1}}, \theta_{a_j}, \theta_{a_{j+1}}, \ldots,$ and if the solutions to (b) are $\ldots, \theta_{b_{j-1}}, \theta_{b_j}, \theta_{b_{j+1}}, \ldots,$ then

$$\cdots < \theta_{a_{j-1}} < \theta_{b_{j-1}} < \theta_{a_j} < \theta_{b_j} < \theta_{a_{j+1}} < \theta_{b_{j+1}} < \cdots.$$

This may seem to be a question that only a mathematician could possibly appreciate, but proving this result was central to establishing several important *engineering* results in the above 1971 paper. There is a certain charm here for the mathematician, too, as this problem is one of those peculiar questions that, despite being quite easy to

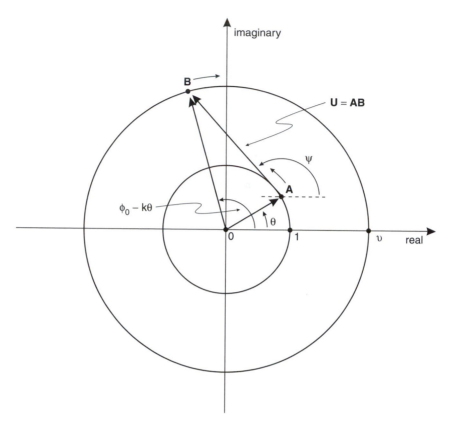

Figure K The *rotating* vector $\mathbf{U} = \mathbf{AB}$

understand, nevertheless frustrates by not presenting a clear place even to *start* an attempt at an analysis. Euler's famous identity, however, will make the proof almost "obvious."

We begin by defining a vector $\mathbf{U}$ in the complex plane, as follows:

$$\mathbf{U} = ve^{i(\phi_0 - k\theta)} - e^{i\theta},$$

which Euler's identity tells us has real and imaginary parts that are precisely the left-hand sides of (a) and (b), respectively. The value of this definition is that it will give us a *geometric* interpretation of (a) and (b), as shown in figure K. In that figure $\mathbf{U}$ is the vector from $\mathbf{A}$ to $\mathbf{B}$, where vector $\mathbf{A}$ is directed onto the unit radius circle (the $e^{i\theta}$ term) and vector **B** is directed onto the (larger) circle with radius v (the $ve^{i(\phi_0 - k\theta)}$ term). The vectors **A** and **B** make angles θ and $\phi_0 - k\theta$, respectively, with the horizontal axis, while **U** (the vector **AB**) makes angle ψ with the horizontal axis. From the definition of **U** we can see that, as θ increases from 0 to 2π, the tip of **A** will rotate *counterclockwise* on its circle (θ's coefficient in **A**'s exponential is *plus* one) while the tip of **B** will rotate in the opposite sense (clockwise) on its circle k times faster than does **A** (θ's coefficient in **B**'s exponential is *minus k*). From the way I've drawn figure K it should be clear that, at least at first, the $\mathbf{U} = \mathbf{AB}$ vector rotates clockwise, that is, angle ψ *decreases* with *increasing* θ. In fact, this will *always* be the case, but, unless your spatial visualization is better than mine, this general statement may *not* be clear. We need to formally establish this assertion, which is the crucial step in the proof of the claim that the roots of (a) and (b) alternate.

What we'll do, mathematically, is show that **U**'s angle ψ with the horizontal axis is, as I stated before, always *decreasing* as θ *increases*, i.e., that $d\psi/d\theta < 0$, *always*. We start by writing

$$\psi = \tan^{-1}\left[\frac{\text{imaginary part of } \mathbf{U}}{\text{real part of } \mathbf{U}}\right] = \tan^{-1}\left[\frac{v\sin(\phi_0 - k\theta) - \sin(\theta)}{v\cos(\phi_0 - k\theta) - \cos(\theta)}\right].$$

Then, applying the standard freshman calculus differentiation formula for the inverse tangent function and doing a little algebra, we find (and you should verify this) that

$$\frac{d\psi}{d\theta} = -\frac{kv^2 - 1 - (k-1)v\cos[(k+1)\theta - \phi_0]}{v^2 + 1 - 2v\cos[(k+1)\theta - \phi_0]}.$$

As θ varies then of course $-1 \leq \cos[(k+1)\theta - \phi_0] \leq 1$. Since $v > 0$, the smallest value for the denominator occurs when $\cos[(k+1)\theta - \phi_0] = +1$, that is, when the denominator is

$$v^2 + 1 - 2v = (v-1)^2.$$

As a squared quantity this smallest value for the denominator is never negative and, in fact, is *positive if* $v \neq 1$.

If $k > 1$, then the factor $(k-1) > 0$ and the smallest value for numerator occurs when $\cos[(k+1)\theta - \phi_0] = +1$, that is, when the numerator is

$$kv^2 - 1 - (k-1)v = kv^2 - kv + v - 1 = kv(v-1) + (v-1)$$
$$= (kv+1)(v-1),$$

which is *positive if* $v > 1$. If $0 < k < 1$, then the factor $(k-1) < 0$ and the smallest value for the numerator occurs when $\cos[(k+1)\theta - \phi_0] = -1$, that is, the numerator's smallest value is now

$$kv^2 - 1 + (k-1)v = kv^2 + kv - v - 1 = kv(v+1) - (v+1)$$
$$= (kv-1)(v+1),$$

which is *positive if* $kv > 1$ (since $k > 0$, by assumption, this condition implies $v > 0$). So, in summary, as long as $k > 0$ then

- the denominator is always positive if $v > 1$;
- the numerator is always positive if $v > 1$ and if $kv > 1$.

Thus, it is certainly true that

$$\frac{d\psi}{d\theta} = -\frac{\text{positive numerator}}{\text{positive denominator}} < 0 \; if \; v > 1 \; and \; kv > 1.$$

This shows that the vector $\mathbf{U} = \mathbf{AB}$ continuously rotates *clockwise* in the complex plane as θ continuously *increases*, if $v > 1$ and $kv > 1$. Now, what can we conclude from this about the roots of (a) and (b)?

Let's arbitrarily start with $\mathbf{U}$ as shown in figure K, pointing into the second quadrant. Both the real and imaginary parts of $\mathbf{U}$ are then *non*zero and $\mathbf{U}$ is complex. As $\mathbf{U}$ rotates clockwise it will eventually be pointing straight up the vertical axis, which means $\mathbf{U}$ is pure imaginary and thus has a zero real part — and so we have arrived at the condition for a root of (a). As the clockwise rotation of $\mathbf{U}$ continues, $\mathbf{U}$ becomes complex again and then, later, $\mathbf{U}$ becomes purely real as $\mathbf{U}$ points along the positive horizontal axis, which means we have a zero imaginary part — and so we have arrived at the condition for a root of (b). Continuing with the clockwise rotation of **U**, you can see that it eventually will be pointing straight *down* the vertical axis (zero real part) and so now we have the condition again for a root of (a). And you can surely at this point see how the argument continues as **U** continues to rotate clockwise — as the potentate of Siam famously says in *The King and I*, "et cetera, et cetera, et cetera."

Notes

Chapter 1: The Classic Pursuit Problem

1. If $f(t) = \int\limits_{h(t)}^{g(t)} K(\alpha, t) \, d\alpha$, then

$$\frac{df}{dt} = \int\limits_{h(t)}^{g(t)} \frac{\partial K}{\partial t} \, d\alpha + K(g(t), t)\frac{dg}{dt} - K(h(t), t)\frac{dh}{dt}.$$

For a freshman calculus derivation of this, see my book *The Science of Radio*, 2nd edition, Springer-Verlag 2001, pp. 416–418.

2. The problem quotation is from Larry C. Andrews, *Introduction to Differential Equations with Boundary Value Problems*, HarperCollins 1991, Problem 22, p. 91. This problem is actually a favorite of textbook writers (recall the quotation from Smith (1917) in *What You Need to Know* . . .), and the very same problem — with the piloted plane replaced with a bird — can be found in a book by two of my former colleagues at Harvey Mudd College, the mathematicians R. L. Borrelli and C. S. Coleman, *Differential Equations: A Modeling Approach*, Prentice-Hall 1987, pp. 112–115. You can find it again in the older (but still outstanding) text by Ralph Palmer Agnew, *Differential Equations*, McGraw-Hill 1960, pp. 146–147. I bought my copy of Agnew new, when a sophomore at Stanford, and it is now — nearly fifty years later — tattered and torn for all the use I've given it over the last five decades.

3. This problem has a superficial resemblance to the *much* more difficult problem of determining the path that minimizes the time required to cross the river. That problem was solved in 1931 by the German mathematician Ernst Zermelo (1871–1953) — and so is called *Zermelo's problem* in the mathematical literature — and it requires mathematics (calculus of variations) beyond the level of this book. You can find an extended discussion of that calculus in my book *When Least is Best*, Princeton 2004, pp. 200–278.

Chapter 2: Pursuit of (Mostly) Maneuvering Targets

1. The "solution" published in the *Diary* goes as follows:

$$\sqrt{2}b = \sqrt{2}\sum_{k=1}^{\infty}\frac{1}{\sqrt{2k}} = \sum_{k=1}^{\infty}\frac{1}{\sqrt{k}} = \frac{1}{\sqrt{1}} + \frac{1}{\sqrt{2}} + \frac{1}{\sqrt{3}} + \frac{1}{\sqrt{4}} + \cdots .$$

Now,

$$a = \sum_{k=0}^{\infty}\frac{1}{\sqrt{2k+1}} = \frac{1}{\sqrt{1}} + \frac{1}{\sqrt{3}} + \frac{1}{\sqrt{5}} + \cdots$$

and

$$b = \sum_{k=1}^{\infty}\frac{1}{\sqrt{2k}} = \frac{1}{\sqrt{2}} + \frac{1}{\sqrt{4}} + \frac{1}{\sqrt{6}} + \cdots .$$

Thus,

$$a + b = \frac{1}{\sqrt{1}} + \frac{1}{\sqrt{2}} + \frac{1}{\sqrt{3}} + \frac{1}{\sqrt{4}} + \frac{1}{\sqrt{5}} + \cdots = \sqrt{2}b,$$

and so $b(\sqrt{2}-1) = a$, or

$$\frac{a}{b} = \sqrt{2} - 1.$$

This is, however, a meaningless calculation as the sums for a and b *individually diverge* (for some discussion on this point, see note 9 for chapter 4). That is what I meant when I warned you of trickery!

2. Ball (1921) also reports that the pursuit problem with a straight line escape path had appeared in the *Ladies' Diary* several years before Ash's problem.

3. The despairing quotes are from two professors of mathematics — see *American Mathematical Monthly*, August 1894, pp. 273–275. The proposer of the problem that prompted those responses was one A. L. Foote, and his problem was later called the "Foote race problem." Who says mathematicians have no sense of humor!

4. Hathaway had earlier played a valuable (if secondary) role in the history of science. Before joining the math faculty at Rose Polytechnic, he had taught at Cornell and, even earlier, at The Johns Hopkins University in Baltimore. While at Johns Hopkins, in October 1884, Hathaway attended a series of lectures given by Sir William Thomson (1824–1907), better known today as Lord Kelvin, while the Scottish scientist was visiting the United States. Apparently all in the audience sat enthralled as Sir William presented his views on molecular

dynamics and the wave theory of light, but Hathaway went one step farther — trained as a stenographer as well as a mathematician, Hathaway kept a verbatim record of the exact words of the great man. In those days before electronic recordings, that was a rare event indeed in the scientific world. Hathaway's transcript provides a true account of Thomson's extemporaneous lecture style, and it is a document still greatly valued by historians. Hathaway's stenographic report of the lectures was later reprinted in unrevised book form under the title *The Baltimore Lectures*, and it proves to be fascinating reading even today.

5. All of the computer-generated plots in this book were produced using MATLAB®*7.3* (the book, itself, was typed on the same computer in Scientific Word®*4.0*). And when I got stuck-in-a-writing-rut, the very same machine smoothly switched over from performing a tremendous number of mathematical calculations per second to let me play (I am ashamed to admit how much) Call of Duty® and other video games — see the Frontispiece for how your author spends a lot of his retirement.

6. To see how ode45 is used in MATLAB®*7.x* ($x = 3$ for this book), see the invaluable reference by D. J. and N. J. Higham, *MATLAB Guide* (Second Edition), SIAM 2005, pp. 175–189. The Highams' book is simply *stuffed* with useful coding examples.

7. Luke Chia-Liu Yuan, "Homing and Navigational Courses of Target-Seeking Devices," *Journal of Applied Physics*, December 1948, pp. 1122–1128. (Notice the interesting euphemism in the title for what most readers surely thought of as being a missile.) Yuan was not a mathematician, but rather was on the physics faculty at Princeton University (he says he actually did the work reported while at the RCA Laboratories in Princeton), and that it was there the discontinuity in missile acceleration at $k = 2$ was discovered. However, in a reply to Yuan, Bacon (1950) writes that "The fact that [the radius of curvature of the missile flight path] approaches zero if k be greater than 2 [and so the acceleration becomes infinite] is well-known to every student of the subject."

8. L. R. Ford, *Differential Equations*, McGraw-Hill 1955, p. 32. The same sort of problem (with analysis) can be found in Agnew (see note 2 for chapter 1), p. 63.

9. This problem is also found in Borrelli and Coleman (see note 2 for chapter 1), p. 116, who use the same numerical values as in Ford's book (note 8). Borrelli and Coleman change the rum runner into a smuggler's launch, but more importantly provide helpful starting hints in the form of a sketch of the expanding spiral path for the coast-guard boat, as well as specifying the initial path of that boat as heading directly toward the launch's last spotted location.

10. In the math puzzle book by C. S. Ogilvy, *Excursions in Mathematics*, Dover 1994 (originally published in 1956), it is stated (p. 148), without proof, that "the spiral must be logarithmic." It would have been better, I think, as you can now see, to say the spiral is *exponential* (although one could argue, I suppose, that since θ increases *logarithmically* with *time* then Ogilvy's statement is okay). Ogilvy, a long-time math professor at Hamilton College in New York, was a well-known expert in sailing and navigation who wrote often on problems of water pursuit. He died in 2000, at age 87.

11. We can set $V_\theta = 0$ without worrying about the possibility that $V_r = 0$, too (which would render (2.6.14) undefined), because there is a pretty little theorem that shows the roots of $V_\theta(\theta) = 0$ and $V_r(\theta) = 0$ must *alternate* along the θ-axis. That is, there is no θ^* such that $V_\theta(\theta^*) = V_r(\theta^*) = 0$. The proof is clever but not difficult — it is a very elegant application of Euler's identity — and you can find that theorem proven in detail in appendix K.

12. Jarrell attached the following explanatory note to his poem:

> A ball turret was a Plexiglas sphere set into the belly of a B-17 or B-24 and inhabited by two .50 caliber machine-guns and one man, a short small man. When this gunner tracked with his machine guns a fighter attacking his bomber from below, he revolved with the turret; hunched upside-down in his little sphere, he looked like the foetus in the womb. The fighters which attacked him were armed with cannon firing explosive shells. The hose was a steam hose.

One detail in the poem may still puzzle — the "wet fur froze." That is a reference to the ball turret gunner's fleece-lined flight jacket, which was barely up to the job of protecting him from the subzero cold at the high altitudes at which the bombers operated. Jarrell's poem was the inspiration for my short science fiction story "The Infinite Plane," *Omni*, April 1981, a stream-of-consciousness tale of a B-17 ball turret gunner who, along with the protective shell of his ball turret, is shot free from his bomber and is falling to what he thinks is certain death into the ocean twenty-five thousand feet below. When I wrote my story I had simply "made-up" the horrific scenario of a gunner trapped in a free-falling ball turret for dramatic purposes, but I've since learned such things actually happened — see *Dead Engine Kids: World War II Diary of John J. Briol, B-17 Ball Turret Gunner*, Silver Wings Aviation 1993, p. 54.

Chapter 3: Cyclic Pursuit

1. H. Bateman, *Differential Equations*, Chelsea 1966 (originally published in London in 1918), pp. 8–10. Bateman, the 1903 Senior Wrangler in the Mathematical Tripos, emigrated to America in 1910. From 1917 until his death he was on the faculty of California's Throop College of Technology (renamed California Institute of Technology in 1920), and his treatment of cyclic pursuit would, decades later, be of great help to analysts.

2. I am of course being facetious when I ask you to "remember" this identity. When I came to it while reading Klamkin and Newman (1971) I had never seen it before, and thought it quite strange-looking. After two long, frustrating days of trying to establish it — and failing — I was applying some strong language to it! This is an interesting example of how what is obvious to one person can be obscure to another, as elsewhere in their paper Klamkin and Newman elaborate on what I thought to be easy steps, and then they simply slip, with *no* explanation, this incredible identity into the text. I am, therefore, most grateful to Professor Eli Maor, who, within *two hours* (!) of my desperate

e-mail to him for help, provided me with an elegant proof. As the author of the equally elegant *Trigonometric Delights*, Princeton 1998, who could have been a better choice for me to turn to for help than Eli?

PROOF OF THE IDENTITY

$b\{\sin(A)\sin(B) - \cos(A)\cos(B) + 1\} = c\{\cos(A) - \cos(B)\} + a\{1 + \cos(C)\}$

Using the identity $\cos(A \pm B) = \cos(A)\cos(B) \mp \sin(A)\sin(B)$ we can write the LHS (left-hand side) of the proposed identity as $b\{1 - \cos(A + B)\}$. Since $C = \pi - (A + B)$, then we can write the RHS (right-hand side) of the proposed identity as

$$c\{\cos(A) - \cos(B)\} + a\{1 + \cos[\pi - (A + B)]\}$$
$$= c\{\cos(A) - \cos(B)\} + a\{1 + \cos(\pi)\cos(A + B) + \sin(\pi)\sin(A + B)\}$$
$$= c\{\cos(A) - \cos(B)\} + a\{1 - \cos(A + B)\}.$$

So, the proposed identity is equivalent to

$$b\{1 - \cos(A + B)\} = c\{\cos(A) - \cos(B)\} + a\{1 - \cos(A + B)\},$$

or, with some rearrangement,

$$(b - a)\{1 - \cos(A + B)\} = c\{\cos(A) - \cos(B)\}.$$

Next, using the half-angle identity $1 - \cos(\theta) = 2\sin^2(\theta/2)$, the LHS of this latest version of the proposed identity becomes $2(b - a)\sin^2((A + B)/2)$. The RHS of this latest version of the proposed identity can be rewritten, using a well-known identity found in any math table of formulas, as

$$c\{\cos(A) - \cos(B)\} = 2c\sin\left(\frac{A + B}{2}\right)\sin\left(\frac{B - A}{2}\right)$$
$$= -2c\sin\left(\frac{A + B}{2}\right)\sin\left(\frac{A - B}{2}\right)$$

and so, after cancelling the common factor of $2\sin((A + B)/2)$, the new version of the proposed identity becomes

$$(b - a)\sin\left(\frac{A + B}{2}\right) = -c\sin\left(\frac{A - B}{2}\right).$$

To finish the proof, we'll now work exclusively with the LHS of the boxed expression and show how to manipulate it to arrive at the RHS of the boxed expression.

We begin by recalling the law of sines, valid for all triangles:

$$\frac{a}{\sin(A)} = \frac{b}{\sin(B)} = \frac{c}{\sin(C)} = \frac{c}{\sin(\pi - (A+B))}$$

$$= \frac{c}{\sin(\pi)\cos(A+B) - \cos(\pi)\sin(A+B)} = \frac{c}{\sin(A+B)}.$$

Thus,

$$a = c\,\frac{\sin(A)}{\sin(A+B)}$$

and

$$b = c\,\frac{\sin(B)}{\sin(A+B)},$$

and so

$$b - a = c\,\frac{\sin(B) - \sin(A)}{\sin(A+B)}.$$

We can now finish our manipulation of the LHS of the boxed expression with a flourish:

$$\text{LHS of the boxed expression} = (b-a)\sin\left(\frac{A+B}{2}\right)$$

$$= c\,\frac{\sin(B) - \sin(A)}{\sin(A+B)}\sin\left(\frac{A+B}{2}\right)$$

or, using another well-known identity for $\sin(B) - \sin(A)$,

$$\text{LHS} = c\,\frac{-2\cos((A+B)/2)\sin((A-B)/2)}{\sin(A+B)}\sin\left(\frac{A+B}{2}\right)$$

or, using the half-angle identity $\sin(\theta) = 2\sin\left(\frac{\theta}{2}\right)\cos\left(\frac{\theta}{2}\right)$,

$$\text{LHS} = \frac{-2c\cos((A+B)/2)\sin((A-B)/2)}{2\sin((A+B)/2)\cos((A+B)/2)}\sin((A+B)/2) = -c\sin\left(\frac{A-B}{2}\right),$$

which is the RHS of the boxed expression and we are done. Thanks, Eli!

3. To establish this double inequality on $f = \cos(A) + \cos(B) + \cos(C)$, under the constraints that $A + B + C = \pi$ and $A, B, C \geq 0$, we can employ the method of Lagrange multipliers. That is, since $g(A, B, C) = \pi - A - B - C = 0$, then finding the extrema of f is the same as finding the extrema of $f - \lambda g$, where λ is any finite constant (λ times zero is zero). Necessary conditions for the extrema of this "modified" f are

$$\frac{\partial(f - \lambda g)}{\partial A} = \frac{\partial(f - \lambda g)}{\partial B} = \frac{\partial(f - \lambda g)}{\partial C} = 0,$$

where we treat A, B, and C as independent. So, since

$$f - \lambda g = \cos(A) + \cos(B) + \cos(C) - \lambda(\pi - A - B - C),$$

we have

$$\frac{\partial(f - \lambda g)}{\partial A} = -\sin(A) + \lambda,$$

$$\frac{\partial(f - \lambda g)}{\partial B} = -\sin(B) + \lambda,$$

$$\frac{\partial(f - \lambda g)}{\partial C} = -\sin(C) + \lambda.$$

For these three partial derivatives to vanish we have

$$\lambda = \sin(A) = \sin(B) = \sin(C).$$

One obvious solution is $A = B = C$ and so $A = B = C = \frac{\pi}{3}$, which gives $f = 3\cos(\pi/3) = \frac{3}{2}$ as an extrema. Another solution is $A = B$ and $C = \pi - A$ because $\sin(A) = \sin(\pi - A)$. Thus, using the constraint $A + B + C = \pi$, we have $A + A + \pi - A = \pi$ or $A = 0$. That is, $A = B = 0$, and $C = \pi$, which gives $f = 2\cos(0) + \cos(\pi) = 2 - 1 = 1$ as an extrema (which is seen to be a *minimum*, while the $\frac{3}{2}$ is a *maximum*). So, $1 \leq \cos(A) + \cos(B) + \cos(C) \leq \frac{3}{2}$ under the constraints $A + B + C = \pi$ and $A, B, C \geq 0$.

Chapter 4: Seven Classic Evasion Problems

1. Solving $\tan(\phi_t) = \pi + \phi_t$ is equivalent to asking for the root of $f(\phi_t) = \tan(\phi_t) - \pi - \phi_t = 0$. This can be easily solved by computer, to any desired degree of accuracy, with the Newton-Raphson method: see any book on numerical analysis, or my book *When Least is Best*, Princeton 2004, pp. 120–123.

2. The plots of figures 4.3.4 through 4.3.7 were created by applying *Euler's method* to (4.3.8) to find Y, and then using (4.3.9) to find X. Euler's method replaces the equation $dY/dT = f(Y, T)$ with the approximation $\Delta Y = f(Y, T)\Delta T$ and calculates an "updated" Y from $Y_{\text{new}} = Y_{\text{old}} + \Delta Y$. This isn't the most sophisticated approach, but it is easy to code and, in using it, I am following in Professor Bailey's footsteps. You can find another detailed example of Euler's method in appendix D, which includes a computer simulation of Morley's "curve of ambience" problem.

3. It is often claimed in modern game theory texts that in 1912 Ernst Zermelo — see note 3 for chapter 1 — proved that chess is *determinate*, that is, that there exist one or more lists (strategies) for either **W** or **B** by which **W** or **B** could *force* a checkmate, or at least a stalemate. That this is not so (while the *assertion* may be true, it is also true that *Zermelo* proved no such thing) is discussed in a fascinating historical paper by Ulrich Schwalbe and Paul Walker, "Zermelo and the Early History of Game Theory," *Games and Economic Behavior*, January 2001, pp. 123–137. The last line of Zermelo's own paper (published in 1913) clearly supports the correctness of the authors' thesis: "The question as to whether the starting position [of a game of chess] is already a winning position for one of the parties is still open. Would it be answered exactly, Chess would of course lose the character of [being] a game at all."

4. Another such attempt at "social mathematics," not as well known as is probability theory, is the mathematical theory of warfare due to the English mathematical aeronatical and mechanical engineer Frederick Lanchester (1868–1946), who modeled modern combat in the form of differential equations. The *Lanchester equations* were put forth in his 1916 book *Aircraft in Warfare*, and they are still an active area of military operations research today.

5. Von Neumann was thought by at least one writer to have been the inspiration for the mad scientist in the 1965 black comedy *Dr. Strangelove or: How I Learned to Stop Worrying and Love the Bomb*. (See, for example, Paul Strathern, *A Brief History of Economic Genius*, Texere 2002.) As Strathern writes in his Prologue, that film posits

> A sole American bomber has eluded the Soviet defense system, and is now beyond recall. With horror, the Soviet ambassador reveals to the President and his advisor [Dr. Strangelove, an ex-Nazi] the likely consequences: if the bomber succeeds in reaching its target, it is liable to trigger the Soviet Union's ultimate [and never revealed] weapon, the Doomsday Machine. This will release a vast cloud of radioactive material which will enshroud the entire Earth, destroying all human and animal life for 100 years. [This grim scenario had already appeared in film, in the shocking 1959 *On the Beach*.] The President is aghast. Dr. Strangelove is exasperated, and exclaims to the Soviet ambassador, 'The whole point of having a Doomsday Machine is lost *if you keep it a secret. Vy* didn't you tell the vurld?'

Strathern then observes, "This is the logic of game theory [that is, both players have *total* information] — the first explicit reference to this new method of strategic thought in a popular movie." This does sort of bring von Neumann to mind, but an equally obvious candidate for Dr. Strangelove might be the real ex-Nazi SS officer Werner von Braun (head of the German V-2 ballistic missile terror weapon program), who after the war headed the American rocket program.

6. Problems like this were among the real-life tasks of the Anti-Submarine Warfare Operations Research Group (ASWORG), formed in 1942 at the beginning of American concern with German submarine activity off of the Atlantic Coast. The story of the start of ASWORG is told by its founding leader in the autobiography of Philip Morse (1903–1985), *In at the Beginnings: A Physicist's Life*, MIT Press 1977. To see what sort of theoretical analyses on the general problem of *searching* ASWORG did during the Second World War, see the famous trilogy of papers written by the American mathematician Bernard Koopman (1900–1981): Koopman (1956a, b, 1957). These papers reported on the unclassified parts of a classified (CONFIDENTIAL) report published shortly after the war (1946). With a broadening of its mission after the war, ASWORG became just OEG (Operations Evaluation Group) and then, in 1962, was merged with the Institute for Naval Studies to form the Center for Naval Analyses (CNA). I worked at CNA as a naval weapons systems analyst (NWSA) during the summer of 1973, long after the glory days of the original ASWORG had passed, but it was nevertheless an interesting experience. To get an idea of what a NWSA *does*, see Marc Mangel, "Applied Mathematicians and Naval Operators," *SIAM Review*, July 1982, pp. 289–300.

7. Isaacs proposed the Princess-and-Monster problem in his 1965 book *Differential Games* as one continuous in both space and time. He was unable to solve it, but suggested that if one could first solve the discrete version then perhaps the continuous case would be the limit as $n \rightarrow \infty$. This was done by Wilson (1972), and that is the work discussed in this book. Unknown to Wilson (who was a graduate student in the Department of Mathematics at the University of Adelaide in Australia), the continuous case solution was published that same year by the Russian mathematician M. I. Zelikin at Moscow State University. Zelikin assumed that both the monster and the princess have independent starting points that can be anywhere in the circular tunnel with a uniform probability distribution. The solution is remarkably simple — see Zelikin (1972) — simple enough to be included, in my opinion, in an *undergraduate* probability course.

8. The simulation results of figure 4.6.3 strongly suggest the theoretical result of (4.6.2) — given without proof — for the average duration of the pursuit. Taking the duration of the pursuit as a discrete-valued random variable, $\mathbf{D}$, the simulation says the values of $\mathbf{D}$ are $D_1 = 5$, $D_2 = 10$, $D_3 = 15$, $D_4 = 20$, ..., that is, are v, $2v$, $3v$, ... (since $v = 5$). The average value of this random variable is, by definition, $\sum_{q=1}^{\infty} D_q \cdot \text{Prob}(\mathbf{D} = D_q)$. From the simulation we can estimate the values of $\text{Prob}(\mathbf{D} = D_q)$, as well as I can read figure 4.6.3, to be

$$\frac{5,000}{10,000} = \frac{1}{2}, \ \frac{2,500}{10,000} = \frac{1}{4}, \ \frac{1,250}{10,000} = \frac{1}{8}, \ \frac{625}{10,000} = \frac{1}{16}, \ \ldots,$$

i.e., $\text{Prob}(\mathbf{D} = D_q) = 1/2^q$. Thus, a 'reasonable guess' for the average value of the duration of the pursuit is

$$\sum_{q=1}^{\infty} \frac{q\,v}{2^q} = \frac{v}{2} + \frac{2v}{4} + \frac{3v}{8} + \cdots$$

which is (4.6.2). Of course, the simulation results are not *quite* perfect; rather than the 'perfect' sequence of numbers $(5,000,\ 2,500,\ 1,250,\ 625,\ldots)$ the simulation produced $(5,006,\ 2,518,\ 1,264,\ 599,\ \ldots)$. These values are still highly suggestive of (4.6.2), however, and the discrepancies (which are not really very large) are the result of the *finite* number of simulations.

9. As is demonstrated in first year calculus, $\sum_{m=1}^{\infty} (1/m^p)$ diverges for $p \leq 1$ (e.g., $p = \frac{3}{4}$) and is finite for $p > 1$ (e.g., $p = \frac{3}{2}$). The borderline case of $p = 1$ is the famous harmonic series which, while divergent — see my book *An Imaginary Tale: The Story of $\sqrt{-1}$*, Princeton 1998, 2006 (corrected edition), pp. 146–147 — goes to infinity at a fantastically slow rate, e.g., the partial sum of the harmonic series doesn't exceed 100 until somewhat more than $1.5 \cdot 10^{43}$ terms have been added!

Bibliography

Aravind, P. K., "A Symmetrical Pursuit Problem on the Sphere and the Hyberbolic Plane," *The Mathematical Gazette* 78, 1994, pp. 30–36.

Archibald, R. C., and Manning, H. P., "Remarks and Historical Notes," *American Mathematical Monthly*, February 1921, pp. 91–97.

Bacon, R. H., "The Pursuit Curve," *Journal of Applied Physics*, October 1950, pp. 1065–1066.

Bailey, H. R., "The Hiding Path," *Mathematics Magazine*, January 1994, pp. 40–44.

Ball, W.W.R., "Curves of Pursuit," *American Mathematical Monthly*, June–July 1921, pp. 278–279.

Barton, J. C., and Eliezer, C. J., "On Pursuit Curves," *The Journal of the Australian Mathematical Society*. Series B41, Applied Mathematics, 2000, pp. 358–371.

Baston, V. J., and Bostock, F. A., "A One-Dimensional Helicopter-Submarine Game," *Naval Research Logistics* 36, 1989, pp. 479–490.

Bateman, H., *Differential Equations*, Chelsea 1966, pp. 8–10.

Behroozi, F., and Gagnon, R., "A Computer-Assisted Study of Pursuit in a Plane," *American Mathematical Monthly*, October 1975, pp. 804–812.

———, "Cyclic Pursuit in a Plane," *Journal of Mathematical Physics*, November 1979, pp. 2212–2216.

Bernhart, A., "Asymptotic Pursuit," *Proceedings of the Oklahoma Academy of Science for 1953*, pp. 164–165.

———, "Curves of Pursuit," *Scripta Mathematica*, September–December 1954, pp. 125–141.

———, "Curves of Pursuit-II," *Scripta Mathematica*, Memorial Issue 1957, pp. 49–65.

———, "Polygons of Pursuit," *Scripta Mathematica*, 1959a, pp. 23–50.

———, "Curves of General Pursuit," *Scripta Mathematica*, 1959b, pp. 189–206.

Bruckstein, A. M., "Why the Ant Trails Look So Straight and Nice," *The Mathematical Intelligencer* 15, 1993, pp. 59–62.

Cady, W. G., "The Circular Tractrix," *American Mathematical Monthly*, December 1965, pp. 1065–1071.

Clapham, J. C., "Playful MIce," *Recreational Mathematics Magazine*, August 1962, pp. 6–7.

Colman, W.J.A., "A Curve of Pursuit," *Bulletin of the Institute of Mathematics and Its Applications*, March 1991, pp. 45–47.

Croft, H. T., "'Lion and Man': A Postscript," *Journal of the London Mathematical Society* 39, 1964, pp. 385–390.

Davis, H., *Introduction to Nonlinear Differential and Integral Equations*, Dover 1962, pp. 113–127.

Dobbie, J. M., "A Two Cell Model of Search for a Moving Target," *Operations Research*, 1974, pp. 74–92.

Eagle, J. N., "The Optimal Search for a Moving Target When the Search Path is Constrained," *Operations Research*, September–October 1984, pp. 1107–1115.

Eliezer, C. J., and Barton, J. C., "Pursuit Curves," *Bulletin of the Institute of Mathematics and Its Applications*, November/December 1992, pp. 182–184.

Finch, S. R., and Wetzel, J. E., "Lost in a Forest," *American Mathematical Monthly*, October 2004, pp. 645–654.

Gluss, B., "The Minimax Path in a Search for a Circle in a Plane," *Naval Research Logistics Quarterly*, December 1961, pp. 357–360.

Guelman, M., "A Qualitative Study of Proportional Navigation," *IEEE Transactions on Aerospace and Electronic Systems*, July 1971, pp. 637–643.

Guha, A., and Biswas, S. K., "On Leonardo da Vinci's Cat and Mouse Problem," *Bulletin of the Institute of Mathematics and Its Applications*, January/February 1994, pp.12–15.

Hackett, F. E., "A Numerical Solution to the Triangular Problem of Pursuit," *The Johns Hopkins University Circular*, July 1908, pp. 135–137.

Handelman, G. H., "Aerodynamic Pursuit Curves for Overhead Attacks," *Journal of the Franklin Institute*, March 1949, pp. 205–221.

Hoenselaers, C., "Chasing Relativistic Rabbits," *General Relativity and Gravitation* 1995 (no. 4), pp. 351–360.

Isbell, J. R., "An Optimal Search Pattern," *Naval Research Logistics Quarterly*, December 1957, pp. 357–359.

———, "Pursuit Around a Hole," *Naval Research Logistics Quarterly*, December 1967, pp. 569–571.

Klamkin, M. S., and Newman, D. J., "Cyclic Pursuit or 'The Three Bugs Problem'," *American Mathematical Monthly*, June–July 1971, pp. 631–639.

Koopman, B. O., "The Theory of Search: Kinematic Bases," *Operations Research*, June 1956a, pp. 324–346.

———, "The Theory of Search: Target Detection," *Operations Research*, June 1956b, pp. 503–531.

———, "The Theory of Search: The Optimum Distribution of Seaching Effort,' *Operations Research*, October 1957, pp. 613–626.

Lotka, A. J., "Families of Curves of Pursuit and Their Isochrones," *American Mathematical Monthly*, October 1928, pp. 421–424.

Luther, H. A., "Solution to Problem 3942," *American Mathematical Monthly*, August–September 1941, pp. 484–485.

Macmillan, R. H., "Curves of Pursuit," *The Mathematical Gazette*, February 1956, pp. 1–4.

Mangel, M., "Search for a Randomly Moving Object," *SIAM Journal on Applied Mathematics*, April 1981, pp. 327–338.

Morley, F., "The Curve of Ambience," *American Journal of Mathematics*, July 1924, pp. 193–200.

Morley, F. V., "A Curve of Pursuit," *American Mathematical Monthly*, February 1921, pp. 54–61.

Murtaugh, S. A., and Criel, H. E., "Fundamentals of Proportional Navigation," *IEEE Spectrum*, December 1966, pp. 75–85.

Musès, C., "De Morgan's Ramanujan: An Incident in Recovering Our Endangered Cultural Memory of Mathematics," *The Mathematical Intelligencer*, 1998 (no. 3), pp. 47–51.

Perl, T., "The *Ladies' Diary or Woman's Almanack*, 1704–1841," *Historia Mathematica*, 1979 (no. 6), pp. 36–53.

Puckette, C. C., "The Curve of Pursuit," *The Mathematical Gazette*, 1953, pp. 256–260.

Raina, D., "Mathematical Foundations of a Cultural Project or Ramchandra's Treatise 'Through the Unsentimentalized Light of Mathematics'," *Historia Mathematica*, November 1992, pp. 371–384.

Rees, M., "The Mathematical Sciences and World War II," *American Mathematical Monthly*, October 1980, pp. 607–621.

Richardson, T. J., "Non-mutual Captures in Cyclic Pursuit," *Annals of Mathematics and Artificial Intelligence* 31, 2001, pp. 127–146.

Ruckle, W. H., "Geometric Games of Search and Ambush," *Mathematics Magazine*, September 1979, pp. 195–206.

———, "A Discrete Search Game," in *Stochastic Games and Related Topics* (T.E.S. Raghavan et al., editors), Kluwer Academic 1991, pp. 29–43.

Schiebinger, L., *The Mind Has No Sex? Woman in the Origins of Modern Science*, Harvard University Press 1989, pp. 41–43.

Schuurman, W., and Lodder, J., "The Beauty, the Beast, and the Pond," *Mathematics Magazine*, March–April 1974, pp. 93–95.

Seery, M. J., "Pursuit and Regular N-gons," *The College Mathematics Journal*, May 1998, pp. 228–229.

Shukla, U. S., and Mahapatra, P. R., "The Proportional Navigation Dilemma — Pure or True?" *IEEE Transactions on Aerospace and Electronic Systems*, March 1990, pp. 382–392.

Smith, D. E., "On the Origin of Certain Typical Problems," *American Mathematical Monthly*, February 1917, pp. 64–71.

Stone, L. O., "Search Theory: A Mathematical Theory for Finding Lost Objects," *Mathematics Magazine*, 1977, pp. 248–256.

Thews, K., "What Are the Size and Shape of the Largest Region That Can Be Guarded Against a Faster Invader?" *American Mathematical Monthly*, August–September 1984, p. 416.

Washburn, A. R., "Search-Evasion Game in a Fixed Region," *Operations Research*, November–December 1980, pp. 1290–1298.

Watton, A., and Kydon, D. W., "Analytical Aspects of the N-Bug Problem," *American Journal of Physics*, February 1969, pp. 220–221.

Wilder, C. E., "A Discussion of a Differential Equation," *American Mathematical Monthly*, January 1931, pp. 17–25.

Williams, J. D., *The Compleat Strategyst: Being a Primer on theTheory of Games of Strategy*, McGraw-Hill 1966.

Wilson, D. J., "Isaacs' Princess and Monster Game on the Circle," *Journal of Optimization Theory and Applications*, 9 1972 (no. 4), pp. 265–288.

Yuan, L.C.-L., "Homing and Navigational Courses of Automatic Target-Seeking Devices," *Journal of Applied Physics*, December 1948, pp. 1122–1128.

Zelikin, M. I., "On a Differential Game With Incomplete Information," *Soviet Mathematics Doklady* 13, 1972, pp. 228–231.

Acknowledgments

Immediately after my retirement in May 2004 from the electrical engineering department of the University of New Hampshire (UNH), my wife Pat and I moved from New England to the beautiful state of Tennessee. We came back to New Hampshire the very next year (we missed our kids), but while in Tennessee I wrote this book. While doing that I received a lot of support from the tolerant waitresses at Donna Akey's Plaza Restaurant in Tellico Village, Loudon, who cheerfully kept my coffee cup full even as I sat, nearly every day (sometimes for hours), scribbling away with my pen on a pile of manuscript pages. (Every now and then I would buy a hamburger, too.) They were all too polite to ask what I was up to, but now they know.

Once I had that pile of paper big enough, I sent it to my long-time math editor at Princeton University Press, Vickie Kearn, who encouraged me to finish the book, and who sweetened the offer with a contract to publish it if I actually did get the job done. During those long, quiet writing days in Tennessee I occasionally found that, while the surroundings were gorgeous, there were no technical libraries available (unless I was willing to risk my life dueling with truckers on Interstate 40 to get to the University of Tennessee Library in Knoxville, 30 miles away). I am therefore grateful to two library friends back at UNH, Professors Barbara Lerch and Emily Poworoznek, who sent me copies of papers I could not have obtained otherwise. Two other librarians who deserve my thanks are Marjorie McNinch and

Barbara Hall, at the Hagley Museum and Library in Wilmington, Delaware. With only the few pathetic clues I could give them to go on, they nonetheless unearthed an obscure document I had searched for without success for years — thanks to them I got a beautiful high-resolution digital image on a CD and you got figure 2.6.3.

After I delivered the book to Princeton, I benefited from two helpful pre-publication reviews; I thank Desmond J. Higham, Professor of Mathematics at the University of Strathclyde, Scotland, and an anonymous reviewer. And once the book was done and "in the works" at Princeton, Jill Harris skillfully guided the book through all the administrative details. The book's copyeditor, Alison Anderson, saved me from not just a few embarrassing errors and missteps.

And finally, I owe *everything* to Pat, whose love and encouragement (and tolerance for my eccentricities — which *are* numerous!) has made it possible for me to write all of my books. I made a lucky catch, indeed, back in that long-ago 1962 pursuit.

Paul J. Nahin
Lee, New Hampshire
June 23, 2006

Index